THE ISLANDS SERIES
MAURITIUS

THE ISLANDS SERIES

Achill
Alderney
†The Aran Islands
The Isle of Arran
The Island of Bute
*Canary Islands: Fuerteventura
*Cape Breton Island
*Corsica
*Cyprus
†Dominica
*The Falkland Islands
Gotland
†Grand Bahama
*Harris and Lewis
†The Isle of Mull
Lundy
The Maltese Islands
†Orkney
*Puerto Rico
St Kilda and Other Hebridean Islands
*The Seychelles
†Shetland
*Singapore
Skye
*The Solomon Islands
*Tasmania
*Vancouver Island

in preparation
Bermuda
Fiji
Guernsey
St Helena
Tobago
The Uists and Barra

---

* Published in the United States by Stackpole
† Published in the United States by David & Charles Inc

The series is distributed in Australia by Wren Publishing Pty Ltd
Melbourne

MAURITIUS

by CAROL WRIGHT

DAVID & CHARLES : NEWTON ABBOT

STACKPOLE BOOKS : HARRISBURG

This edition first published in 1974
in Great Britain by
David & Charles (Holdings) Limited Newton Abbot Devon
in the United States in 1974 by
Stackpole Books Harrisburg Pa

7505

ISBN 0 7153 6503 7 (*Great Britain*)
ISBN 0 8117 0994 9 (*United States*)

© CAROL WRIGHT 1974

969.82 T

Æ VIC-L

*Set in eleven on thirteen point Baskerville
and printed in Great Britain
by Latimer Trend & Company Ltd*

CONTENTS

ILLUSTRATIONS

7

ILLUSTRATIONS

Photographs not acknowledged above are from the author's collection

IN TEXT

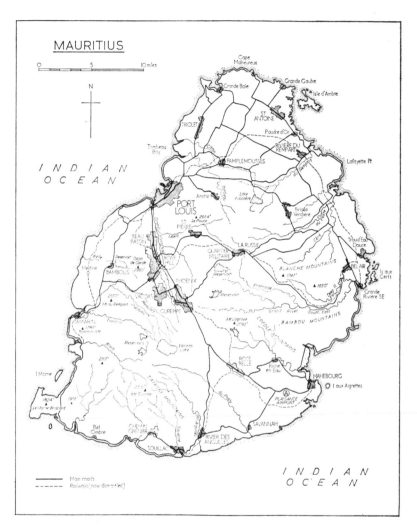

GENERAL MAP OF MAURITIUS

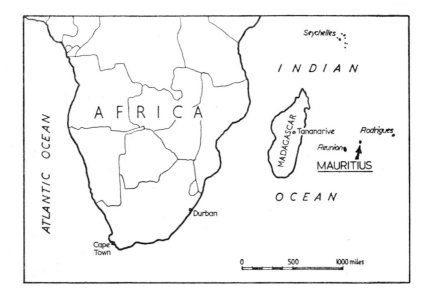

LOCATION OF MAURITIUS IN THE INDIAN OCEAN

1   INTRODUCTION

MANY people think of Mauritius only as the home of the dodo, an early victim of man's pollution of his environment, or for the issue in 1847 of the now immensely valuable penny red and twopenny blue stamps. The Mauritians treasure without malice a letter from Queen Alexandra addressed to 'Mauritius, West Indies', though the island is 500 miles east of Madagascar and 1,250 miles off the South African coast, lying almost on the Tropic of Capricorn. To the east there is nothing except the dependent island of Rodrigues, 360 miles away, to the south nothing but Antarctica and northwards in the Indian Ocean the nearest land is the Seychelles Islands, which were once linked with Mauritius under the British crown.

Her small size—38 miles long by 29 miles wide and roughly pear-shaped—and remote position in the Indian Ocean made Mauritius a late starter in human historic terms. Created by sea volcanic activity in a series of eruptions after Rodrigues was formed, the surface became totally covered with thick forests. The processes of erosion wrought the mountain lava plugs, river valleys and gorges into weird dramatic shapes. Birds were the main indigenous wild life with two species of snakes on Round Island off the north shore. Nicholas Pike, the American consul in Mauritius, writing in the 1870s describes the general appearance of the island, and his description still holds today:

> The physical aspect of the Island is in general picturesque, from the bold and grand outlines of the lofty hills, with their peculiarly

11

formed and varied summits. The north is, for the most part, an elevated plateau, rising to above 1500 feet beyond sea-level. From this elevation, the principal mountain ranges diverge, and the land descends gradually from Curepipe to Grand Port. The eastern side presents a rich and well-cultivated district.

The coast is deeply indented with bays; but there are only three safe approaches for vessels—the Harbour of Port Louis, the Bay of Grand Port and the Baie de la Rivière Noire.

Mauritius is indeed a beautiful island. Her beaches, particularly on the east coast, display all the white ground coral sand, palms and feathery casuarina trees that the tourist brochures could dream of. In her central areas are toothy, deeply caved mountains, so eroded and nature-sculpted that one current Mauritian writer, Malcolm de Chazal, theorises that the mountains were 'built' by the giants of Limuria and not, as the geologists tell us, shaped by the action of the elements. The mountains are remnants of a much older island; in the strata of Le Morne mountain, for instance, it is easy to trace the projections that show how large it once was.

One of the best views of the mountains is to be seen when entering Port Louis harbour; they rise in terraces from the semicircle of city at sea level. Their beauty has been described by various visitors, who were no doubt as much relieved at reaching the island without being dashed by winds on to the reefs as appreciative of the scene. Nicholas Pike arrived from America in January 1867. 'On nearing the land, the fields of waving canes, topes of cocoas and groves and casuarinas, gave a pleasing impression of the place; but when approaching Port Louis harbour, the beauty of the view is unsurpassed and no easy task to describe.' He tries, however, listing the mountains, which 'all formed an entourage few cities can boast, and rendered it when viewed from the sea, the most picturesque in the world'.

Darwin was more of a globetrotter than Pike, and though he praised Mauritius lavishly, spoke of the scenery as 'intermediate in character between the Galapagos and Tahiti but without the charm of Tahiti or the grandeur of Brazil'. He was, however, much taken with the vistas of mountains, which impress every visitor: 'Their summits, as so commonly happens with ancient volcanic rocks, being jagged into the sharpest points. Manes of white clouds were collected round these pinnacles as if for the sake of pleasing the stranger's eye.'

Behind the ground coral beaches lie river gorges and gullies broken by cascades, rising to the high central plateau and mountain ranges. Pike described the rapid rise and fall of the rivers as follows: 'There are no less than sixty rivers and streams flowing to the sea, but all are small; very many cease to flow in dry weather, and the largest are only full after heavy rains when their rise is so rapid as often to occasion much mischief— but they descend to their ordinary level with equal rapidity.' Of the mountains he says: 'Those near the coast are mostly rugged and barren while the mountains towards the interior of the island are well wooded and of great interest to the naturalist. The highest mountain is the Piton de Rivière Noire at 2711 feet, though coming upon the mountains at sea level the effect is much more dramatic and towering.'

This enchanting place was, until the end of the sixteenth century, uninhabited by man. Now it is overpopulated and birth-control programmes are run by the government, though their appearance on TV is resented by the somewhat conservative and puritanical society of Mauritius. The seething mixed mass of people crammed 1,000 to each square mile (a total estimated population in December 1971 of 830,606) strikes most visitors forcibly when they arrive by sea at Port Louis. The crush on the jetty provides an instant kaleidoscope of race. The European French and few British are greatly outnumbered by Indians, Muslims, Malagasy creoles, descended from early sugar plantation slaves, and Chinese. The present mixture has been the same for over 100 years, the Indians and Chinese coming in to replace creole labour after the abolition of slavery in the island. In 1867 Pike reported on 'creoles and coolies,

13

Arabs, Cinghalese, Malagash, Chinese and Malabars; all as eager as in other parts of the world to take the stranger in and carry him off, body and baggage to the nearest hotel'. Darwin, too, was impressed: 'The various races of men walking in the streets afford the most interesting spectacle in Port Louis. Convicts from India are banished here for life; at present there are about 800 and they are employed in various public works.'

Though it is fascinating to watch such racially intermingled groups of people walking the narrow hot city streets, the over-population leads to feelings of oppression and the wish to escape. On seemingly deserted beaches men and boys appear from behind the casuarinas, where they have been fishing or loafing; the cane in the country hides the workers, but pause for a moment of peace and people emerge to watch you. The many small islands round the coast are short-term escapes for the young at weekends, while others take out fishing boats way beyond the confines of the reefs. Crowds wait on the Port Louis dockside or Plaisance airport for glimpses of the outside world arriving and departing. The wealthy and brainy can get away to the mecca of Europe, but the less fortunate have to stay. Many without jobs take the 'relief' work that is organised, and girls become prostitutes if there is no work in the fields or in domestic service. The heat, and isolation from the world, reduce much of the population to apathetic indifference.

Whatever their colour or creed, the lifeline of the island's population is sugar. Of Mauritius' 460,000 acres, over half is under sugar cane, the crop which survives the cyclones. With the remains of forests, the palm groups—sadly now decimated by the rhinoceros beetle—and the rich flowering shrubs, coupled with a reasonable rainfall throughout the year, Mauritius presents a green lush aspect at first glance. The green cane waves against the background of brown-blue lava mountains, which in the bright clear sun play tricks with the eyesight, alternately advancing and retreating at each bend in the road; and the edging of reef water curls languid and turquoise around this greenness. The reefs are studded with small islands to the south-west, south and east, and in the north are the larger satellites—Flat Island, Coin de Mire, Serpent and Round Island, the last now designated a national nature reserve.

In short, Mauritius has an idyllic setting. Darwin said that 'the whole island, with its sloping border and central mountains, was adorned with an air of perfect elegance: the scenery, if I may use such an expression, appeared to the sight harmonious'. Alas, this Garden of Eden setting often has its harmony disturbed by cyclones! The sub-tropical maritime climate is divided into two seasons—the winter from June to October and summer from November to April. The transitional periods between the two seasons—April to June and September to November (roughly)—are pleasant. The climate is not extreme. In summer maximum day and night temperatures in Port Louis are around 31° C and 24° C respectively, and in winter averages are 25° C and 20° C. The highest recorded temperature is 36° C, at Port Louis, and the lowest 7° C, at Curepipe.

Though Mauritians often experience drought, the rainfall and large lake reservoirs and dams over the island give large quantities of water. Rain falls mainly in summer. At sea around Mauritius the annual rainfall is about 40in per annum, but the mountains cause an uplift of moisture-laden air from the sea, resulting in an annual rainfall of almost 60in on the south-east coast and 200in on the central plateau. Much of this rainwater runs straight into the sea, and there are underground streams and freshwater channels that appear even beneath the sea surface, as at Trou d'Eau Douce in the north-east of the island.

During January, February and March the cyclones are at their worst, the most destructive striking about once every 15 years. The last really severe one was cyclone 'Carol' in 1962. Cyclone 'Fabienne' in 1972 avoided Mauritius but caused much damage in Rodrigues. The islanders accept cyclones with practical stoicism: 'class 4' warnings are given on the radio, the people retire to their houses or cyclone shelters, the ships leave harbour to ride out the storm well beyond the reef, and animals are driven into pens. Then come vast winds, sweeping flat and fierce out of the horizon over the pounded reefs, wrecking and flattening. After they have passed, the fields are tidied up, the rivers dredged of debris, trees lifted off the road, live electric wires tied up out of harm's way again and with insurance compensation buildings are replaced. One rebuilt shop in the north-east was renamed 'Magazin Carol' after the 1962 cyclone that blew it down.

Page 17 Port Louis: (above) the harbour showing the ring of hills surrounding the town in the centre of which is the Pouce, and to the left the Pieter Both can also be seen; (below) the statue of Mahé de Labourdonnais, the founder of Port Louis, gazes out on the harbour he created. In the background, down the long palm avenue, is Government House which he also built

Page 18 The sea: (*above*) Ile de la Passe lighthouse built by the French at the entrance to Grand Port harbour and opposite Mahebourg. This was the site of the great naval battle between the French and English in 1810; (*below*) wrecks are common round the Mauritius coast caused by reefs and cyclones. This one, the *Tayeb*, breaks up on the reef inside Port Louis harbour, the victim of a February 1972 cyclone

2   HISTORY

BEFORE 1598 there were no human settlements, though
Phoenician, Malay and Arab sailors knew of the island,
lying as it did on the direct route from Malaya to
Madagascar, and used it for watering on their voyages. The
dodo and other flightless birds, ignorant of predators, roamed
the shores. The aphanapteryx had long legs and escaped the
sailors more easily than the puffy waddling dodo; but once
man started to come regularly, as did the Portuguese, and
deposited pigs, goats, cattle, dogs and rats on the island—the
last two involuntarily, the rest brought from Madagascar in
the hope that they would multiply and provide food for passing
sailors—the birds were doomed. The rats helped destroy the
birds' food and ate their eggs in the ground-level nests.

THE PORTUGUESE DISCOVERIES, 1510–39

The Arabs called Mauritius Dinarobin (or Divarobin) on their
maps, but the Portuguese, who came several times between
1510 and 1539, were the first to explore the island. Pero
Mascarenhas discovered the island of Reunion about 1512 and
after him the three islands in this area (Reunion, Mauritius
and Rodrigues) are called the Mascarenes. Mauritius was
discovered by Domingo Fernandez about 1511 and Rodrigues
by Diego Rodrigues in 1538. Although the islands were on the
Portuguese route to the Indies, no settlements were created.
The Portuguese called Mauritius Ilha do Cerne (Island of the
Swan), a name commemorated in one of the leading newspapers,

B   19

MAURITIUS

Le Cerneen. Some say the sailors had mistaken the dodos on the shores for swans.

THE DUTCH, 1598–1710

The Portuguese did not stay and it was not until 1598 that the island received any settlers, the Dutch, who kept a small community, with slaves, on the island till 1710, when they finally abandoned it. In spite of Dutch governors' attempts to persuade their superiors in Cape Town and Java that Mauritius (named by them after Prince Maurice of Nassau) was worth developing into a boatbuilding centre and a base in the Indian Ocean, only a handful of settlers came, and the odds against even a tolerable life were heavy. Cyclones, then as now, battered the island, ruining the crops; and disease and shortage of food followed. Ill-treated slaves ran away to the forests, where they hid and attacked anyone pursuing them. The climate was hot and unhealthy and ships were often wrecked. An early Dutch governor, Pieter Both, was drowned off the west coast in 1615; he is now commemorated in the name of one of the island's most spectacular mountains.

The Dutch under Admiral Wybrant Van Waerwick first landed at Vieux Grand Port in 1598 near the present Mahebourg, where a few ruins of their dwellings can be seen. They managed to map the island, and chose the big, if windswept, Grand Port as their main settlement, rather than what is now Port Louis, the capital, which they called North West Port. In 1642, Tasman, after mapping the island roughly, left Mauritius on his voyage of discovery to Western Australia. After the arrival of the Dutch the island attracted more notice. French and English ships came, and Thomas Herbert in 1629 noted Mauritius on his travels as 'an island paradise'. He hunted wild hens with a red handkerchief. The Earl of Southampton and Charles I considered the possibility of colonising the island but did not want to provoke war with the Dutch.

The Dutch have been blamed for much unhappiness in Mauritius. By 1680 the dodos had probably disappeared, and the Dutch were also credited with having wiped out the vast ebony forests that covered the island. In fact, although they busily chopped down trees to make roads through the island—for example, north to Flacq, a name still commemorating the Dutch period—there were still plenty left for the French to build their ships and houses. Such wooden houses in Port Louis lasted 200 years or more. Maps even as late as the 1830s show the major part of the island still covered with forests.

Trying to overcome shortages of food, the Dutch imported sambhur deer from Java in 1639 and these increased successfully. They also introduced what was to prove the most important single factor in the island's life, the sugar cane, though little was produced until the middle of the eighteenth century under the French.

The Dutch set up their first permanent settlement in 1638, importing slaves from Madagascar and convicts from Java, but the colony developed little and by 1652 the Cape of Good Hope was a better port of call than Mauritius. Although there were further attempts to colonise, the struggle with nature drove away most settlers.

THE FRENCH, 1715–1810

Deprivation had driven out the remaining handful of settlers by 1710 and the island was abandoned to deer and runaway slaves; from the safety of the forests the latter harried the crews of ships seeking harbour, water, or food. In 1715 the French East India Company claimed the island for France in the name of Louis XIV and sent settlers, the first arriving at the end of 1721. Both Reunion (then called Bourbon) and Mauritius (now rechristened Ile de France) were made the private property of the French East India Company. In 1735 Bertrand François Mahé, Comte de Labourdonnais, was appointed governor.

MAURITIUS

A tough practical Breton sailor who had gone to sea at the age of ten, Labourdonnais' rule from 1735 to 1746 is considered to have been the true foundation and consolidation of the colony of Mauritius. He began by following a decision taken by his predecessor, Nicolas de Maupin, and moved his headquarters from the site of the Dutch settlement to Port Louis in the northwest, where he build Government House and set up boatbuilding yards in the inner harbour. Labourdonnais wanted to create a naval base that would rival that of the British in Bombay, and in fact the activities of French privateers based in Mauritius and attacks by French ships proved a check to the dominance of the British in the Indian Ocean. Labourdonnais received little support from his superiors, who resented their ships being used for fighting rather than commerce, and in 1767 the French East India Company, ruined by wars, sold the island to the French Government.

While he was governor, Labourdonnais created a good anchorage at Port Louis, regulated the food supply and, above all, established internal law and order. He encouraged sugar production, instituting the first mills at Ville Blague and on his own estate at Mon Plaisir. Sturdy stone barracks and fortifications were built at Port Louis, and forest sections were let to contractors who engaged to supply timber for the fleets. The population, particularly the farmers, were much harassed by maroons—runaway slaves—in the forests. In 1736 Labourdonnais recruited a detachment of skilled forest hunters from Bourbon to train soldiers to fight in the interior, and by 1740 could report that only 20 male maroons remained at large.

Alexander Dalrymple of the English East India Company, writing in his journal, summed up the contribution Labourdonnais made to the history of Mauritius and its future effect on British Indian ocean expansion:

From a wild and almost desert island, Mahé de Labourdonnais produced a flourishing and profitable settlement to the French

Company and I am much afraid that port will give them greater advantage over us in India than we at present seem sensible of.

The new Ile de France flourished mainly on account of its geographical position, Port Louis becoming a haven for all shipping in the area. The harbour was reported to be a forest of masts, and even massive cyclones, which in the eighteenth century twice threw many ships up on the shore, did not diminish its importance. Pirates from Madagascar settled in Reunion and Mauritius, and slaves were brought from Madagascar to the sugar plantations.

THE PIRATES

The settlement of the West Indies at the end of the seventeenth century drove many boisterous buccaneers across the world to Sierra Leone and then to Madagascar, where they set up their own republic about 1685. Called Libertalia, it lasted until 1730 as a communistic state under English and French leaders. All Indian Ocean shipping was its prey, though the British India-men suffered particularly. The rich booty captured from these laden merchant ships and taken to Libertalia was eagerly sought after by American traders, who were not above con-doning piracy at the expense of their British colonial masters.

When Madagascar was colonised by the French, the pirates began to drift to Reunion, to help in the settlement there, and to Mauritius. Between 1710 and 1721 the pirates had been using Mauritius as a base. Back in 1702 and 1704 John Bowen and his men had stayed there, to the dismay of the Dutch settlers, though there is no evidence of any fighting between them. At the beginning of the eighteenth century Denmark and others showed great interest in developing the Indian Ocean. In 1720 a vessel was even sent from Ostend to Mauritius to claim the island in the name of the Emperor of Austria. In 1721, however, the island came firmly under the control of the French East India Company.

The French were the first to practise privateering in the Indian Ocean, from the end of the sixteenth century, and Labourdonnais has been described by Mauritian historian Auguste Toussaint as 'first of all a privateer, a "commando" man and not an orthodox strategist like Suffren' (the French commander of the fleet in the area). In the seventeenth century the pirates had been freebooting outlaws, but in the eighteenth, for many of them, their piracy became legalised under the name of privateering. Business could still come before patriotism. There was collusion between English merchants in Madras and the Mauritius pirates, and the resulting pile up of plundered goods in the island led to the rise of a merchant class in Mauritius.

The first privateering expedition from Mauritius took place in the 1740s, during the War of the Austrian Succession. In the Seven Years War, 10 years later, twelve more expeditions under D'Estaing and de la Palliere harried English settlements. In 1762 de la Palliere captured seven Indiamen worth 2,500,000 livres tournois. The rich and riotous living of the pirate era was reflected in the Port Louis of 1772, where over 100 illegal houses sold brandy and sugar until Pierre Poivre had them closed.

The American Revolutionary War and the wars with France encouraged the spread of piracy or privateering. Between 1793 and 1802 prizes taken from the English were estimated at £2,500,000. The archives in Port Louis record twenty-nine privateering expeditions between 1779 and 1782; in 1781 alone there were fifteen. At this time the most celebrated pirate was Deschiers Kerulvay, who managed to collect 1,670,000 livres tournois in a voyage to Ceylon. During the Napoleonic Wars the activity of pirates from Port Louis was as great as that from St Malo in Brittany, and many of the pirates and Mauritian sailors, from Labourdonnais on, had family links with this Breton town.

At this time Robert Surcouf won the title of King of the

Pirates, his bold defiance of governments giving him prominence, though his fellows, such as Malroux, Le Vaillant, Trehouart and Dutertre, were his equals in dash and daring. Toussaint thinks the most audacious of them all was Ripaud de Montaudevert.

The loot poured into Port Louis. Its markets became overstocked, and the American traders received better terms than they could have got in Calcutta. The first American ships came to Mauritius in 1786 for oriental products. Shipowners in the newly formed United States had developed fast clippers and needed a port in the Indian Ocean from which to trade with China. The years 1804–5 saw the peak of American and Mauritian trade, but in 1807 President Thomas Jefferson ended the American voyages. The cessation of this lucrative trade with the USA, Toussaint thinks, contributed to the French loss of their colonies.

Though some English merchants connived at it, it seems doubtful whether the French-inspired piracy could have overthrown the British supremacy in the Indian Ocean. But the Mauritian piracy certainly held the British in check during the eighteenth century, and probably drove them to take over the island. If Labourdonnais' dream of good shipbuilding yards in Port Louis has been realised, more could have been done to defeat the British, but the pirates were using small barques against armed traders. In the Napoleonic Wars the French commander had first-class ships to use against the British in the Indian Ocean, but he quarrelled with Governor Decaen of Mauritius and went back to France with the best of them. In 1806 the British captured the Cape of Good Hope, isolating the Mascarene Islands from European aid.

The Mascarenes had always been badly supplied by France, and therefore turned to America for naval supplies in exchange for British pirated goods. In 1799 the United States signed a private treaty of non-aggression with them. The islands were at that time opposed to the ruling French revolutionary govern-

25

ment on the issue of the abolition of slavery. Decaen, cut off from other help, was forced to commandeer pirate ships, including Surcouf's. The pirate king was so angry at this that he too returned to France. From 1806 to 1810 Decaen had only six frigates while the British were eliminating the privateers by a tight blockade round Mauritius. The French islands, however, managed to keep going on provisions supplied from contraband by the Dutch at Tranquebar.

THE BRITISH, 1810–1968

Though the cosmopolitan East India Company was so important that even its rivals could not do without it, and it was to the interest of all Europeans to maintain the British in the Indian Ocean, intrigue between Decaen and the Indian princes stirred Britain finally to deal with Mauritius. Toussaint thinks it was the intrigues in India rather than the harassing of her ships that finally forced Britain's hand. The British had carried out some naval reconnaisance of the island as early as 1748 under Admiral Boscawen but no attack had been made.

In the Napoleonic Wars Britain blockaded Mauritius, but war was not so total in those days. It is said that when the popular Governor Malartic died of apoplexy, the British officers commanding the naval blockade round Mauritius attended his funeral. Rodrigues was already occupied by the British. Despite a naval defeat for the British at the battle of Grand Port on 24 August 1810, General John Abercrombie and his men were able to land in the north of the island near Cap Malheureux in December 1810. After 4 days Governor Decaen surrendered the island with little bloodshed—150 British killed or wounded—and no pitched battle; indeed the French garrison of Port Louis was not even taken prisoner. General Abercrombie wrote enthusiastically of 'this fine and valuable colony . . . a very fine island and the port establishments of Port Louis are upon a scale of grandeur and magnificence of which I would have formed no idea. The harbour is good and I should imagine must be secure against even the most violent hurricanes'. The general was wrong in this, and not all the new overlords loved Mauritius so unreservedly. Colonel C. G. Gordon (Gordon of Khartoum) who served in Mauritius in 1881 wrote to Sir Charles Elphinstone:

> It is only fair to let you know what you escaped. They say that HRH in one of his furies with someone said to the Adjutant General 'send him to hell'. To that the Adjutant General said 'We have no station there, your Royal Highness'. On which HRH said 'send him to Mauritius' . . . This place is utterly without defence. The sole defences are two 6½-ton guns and yet the War and other Ministers are perfectly content and wrapped up in the ignorance as happy as lambs . . . I have gone into the whole question of defence and put up a report but I suppose it will be considered facetious. If you want to find a place where things have been let go to sleep I recommend you to try Mauritius.

Maybe, with the nineteenth-century supremacy of the British in the Indian Ocean, Mauritius was militarily dormant, but intellectually and commercially she remained lively until the unusually serious malaria outbreak of 1867 and the opening of the Suez Canal 2 years later diminished both her strength and her wealth.

By the Treaty of Paris in 1814 Britain had returned Reunion to France but was ceded Mauritius and its then dependencies—Rodrigues and the Seychelles. The British left French culture and customs as they were and merely changed some names: Port Napoleon reverted to Port Louis, Port Imperial to Grand Port and the Ile de France was restored to the Dutch name of Mauritius. Darwin some 20 years after the arrival of the British noted: 'Although the island has been so many years under the English government, the general character of the place is quite French: Englishmen speak to their servants in French; indeed I should think that Calais or Boulogne was much more Anglified [than Port Louis].' By 1850 Port Louis was prospering like 'a beehive in blossom time' in spite of epidemics like the one in 1844, which killed 12,000 cattle and 6,000 pigs, and attacked other animals. Camels had to be brought to Mauritius to help with gathering the crops.

Prosperity declined more through man than natural disasters. Historian Auguste Toussaint considers the decline of Port Louis was partly due to local greediness, which forced prices up, and to outdated quarantine laws, which kept ships hanging about too long at a time when Mauritius could have become the entrepôt between Europe and Australia. A general high-handedness and soaring prices led ships to avoid Port Louis whenever possible. Nicholas Pike wrote in 1870: 'The whole system of the customs, port dues and in fact all connected with the shipping is calculated to prevent ships entering. No appeals are of any effect to get fair and liberal arrangements and the decline in the shipping tells its own tale. Formerly nearly all our [American] large fleet of whalemen put in here, and left an enormous sum annually for supplies.' The heavy port dues caused whalers to put in instead at Reunion or the Cape of Good Hope. Toussaint agrees with this and thinks Mauritians should have gone in for the lucrative whaling instead of letting it be dominated by the Americans; he feels commercial fishing in general could well have been developed by the island. The waters around the St Brandon and Rodrigues dependencies of Mauritius were rich in fish. Pike states: 'Our American whalemen cruise constantly in the waters near these islands, and numbers of vessels are annually laden with the spoils of the monster sperm whales found in this vicinity.'

Reunion had no natural harbours and depended on Mauritius as her carrier, but the high price of trading through Port Louis aroused her so much that as early as 1799 she threatened to proclaim her independence from France and side with the British. Malartic, then governor of Mauritius, had to go to Reunion to persuade the islanders to change their minds. The tension increased under Governor Decaen. Since 1814, when

Reunion was returned to France by Britain, Mauritians have bought their French goods from Reunion. The Seychelles were then bound with Mauritius as a colony, and likewise suffered the strain of high port prices until they achieved separate financial government in 1872. Mauritius' dependence on Madagascar for her beef has kept tensions to a minimum between the two islands.

Joseph Conrad, who called at Port Louis when he was Captain Korzeniowski, complained of long waits for ships' supplies and of high prices. He paints a portrait in his short story 'A Smile of Fortune' of the sleek rich Port Louis merchant forcing unwanted goods on the captain. Conrad resigned his commission rather than return. He also sketches the other ruling class in Port Louis:

> . . . the old French families, descendants of the old colonists; all noble, all impoverished and living a narrow domestic life in dull, dignified decay. The men, as a rule, occupy inferior posts in Government offices or in business houses. The girls are almost always pretty ignorant of the world, kind and agreeable and generally bilingual; they prattle innocently in both French and English. The emptiness of their existence passes belief.

In the heat men wore correct fustian stiff clothes and rode in carriages, and the women wore copies of the latest French fashions; music, entertaining and dancing were common pastimes.

Mauritius may have been just a dot in the ocean to Europeans, yet in the late eighteenth and early nineteenth centuries it attracted writers, philosophers, travellers and geographers. In 1763 the Abbe de la Caille conducted the first proper survey and mapping of the island, naming many places after people who worked with him. The French explorer La Perouse visited it several times between 1772 and 1777, and met his future wife there. Bernadin de Saint Pierre, author of *Paul et Virginie*, was a royal engineer on Mauritius from 1768 to 1770. Bory de Saint Vincent and Matthew Flinders came and saw and recorded,

Flinders less happily, since he was detained on Mauritius for 6 years during the Napoleonic Wars on his return from a survey in Tasmania. Charles Darwin and the *Beagle* arrived in Port Louis on 29 April 1836 for a 10 day stay.

Darwin, unlike Gordon, loved Mauritius. 'How pleasant it would be to pass one's life in such quiet abodes', he wrote. He stayed with a friend called Lloyd who owned an elephant, and Darwin rode it round in the island. One of the best of the many descriptions of the island was written by Colonel Nicholas Pike, the US consul in the 1860s. In 1873 Pike published his book *Sub Tropical Rambles in the Land of the Aphanapteryx*, in which he describes many of the natural phenomena round the island which the visitor today can still enjoy.

The enquiring minds of the early visitors from Europe aroused in Mauritians a fierce introspective passion which exists today. The islanders are in love with their own history; they will spend hours debating a point of historical fact, and write endlessly in the papers about the past. They still nourish the Mauritius Institute, founded in 1880, and the more recent Historical Society, whose members mark the points throughout the island where events of importance occurred with simple stone cairns. The sense of continuity is strong. The same family names dominate the scene now as in the eighteenth and nineteenth centuries. The archives are intact from 1721. Apart from cyclones and malaria there has been little to change the life of this island. Probably the most significant factors have been the growth of the sugar trade and the end of slavery.

The French Revolution was accepted fairly calmly on the island, with parties, meetings, the setting up of an unused guillotine, and a breaking out of a rash of tricolour ribbon decorations on the ladies' clothes. But once the Revolutionary government's representatives arrived with instructions to abolish slavery, the sugar estate owners promptly hustled them back on board their ships—a local custom with bearers of unwelcome news—and the island remained in a curious suspended outlawry from France until Napoleon's reign.

Once the British were settled in they set about abolishing slavery, an act which seemed to cause more resentment and upheaval than their arrival itself. It was accomplished in 1835 after payment of £2 million in compensation to slave owners. The freed slaves refused to work any more in the cane fields, so the government was forced to bring in coolie labour from China and from India, which began to change the balance of culture and population in the island.

From 1825 the island flourished with exports of sugar to Britain which nearly trebled in the decade 1823–33. With the influx of Indian labour less food was grown locally and rice imported from India. In 1886 a Council of Government was created and elected members included in this.

At the beginning of the twentieth century, the island recovered slowly from a series of epidemics and cyclones. In 1908 Sir Ronald Ross visited Mauritius to begin a twenty year programme of replanning sanitation, water provision and anti-malarial campaigns. Mauritius received grants of aid from Britain and a graduated poll tax on incomes was replaced by income tax in 1951.

After the 1914–18 war there was a period of prosperity due to the boom in sugar prices. At this time a number of minor industries were developed including tea, brick and tile making, aloe fibre bags and a government dairy.

During World War II more local crops were produced since supplies of rice from India and Burma were cut off. The island assumed strategic importance through the closing of the Suez canal and the threat to India by the Japanese. Mauritius organised her own local defence and sent many men to serve overseas.

In 1968 Mauritius was granted independence; it became a member of the Commonwealth, with the Queen's representative as governor general. It receives defence grants from Britain,

31

which maintains a telecommunications station, HMS *Mauritius*, on the island.

The coat of arms of Mauritius, granted in 1906, presents a composite picture of the island. A dodo faces a deer, both holding sugar cane. Palms are laced over them and the motto is 'The Star and the Key of the Indian Ocean' (*Stella clavisque maris India*). Now the dodo is extinct, the deer are being exploited as meat, sugar cane is still the main crop, and the palms are being attacked by the rhinoceros beetle. As a governor of Mauritius once remarked: 'The star is tarnished and the key rusty.' Mauritius is certainly a physical star in her beauty, but with the decline in the importance of shipping she is no longer an ocean key. In her strange mixture of populations, however, she may well prove an exemplary passkey to future human relations.

3 GEOLOGY, FLORA AND FAUNA

BEGINNINGS

'IKE most other islands in the Indian Ocean, the Isle of
France is of volcanic production. Endless are the peculiar
characteristics of its mountain peaks and the abrupt
gigantic fissures which separate them, and of the beds of lava
of different thicknesses and nature which are found everywhere',
wrote Nicholas Pike in 1867. Maurice Tyack, the Mauritian
writer, summed up the geological emergence of Mauritius as
'this fragment of the sunken earth brought to light'.

The Indian Ocean is made up of two great marine basins
separated by a sunken ridge running southwards from India.
Mauritius is a visible part of this submarine ridge, probably
the worn down top of an immense shield volcano which was
originally built up from the sea bed. The island is entirely
volcanic, except for small marginal tracts of uplifted coral reef
and beach material. It is believed that there were two separate
phases of volcanic activity and in between a long period of
erosion. The island was in existence by the beginning of the
Pleistocene epoch. No traces of important mineral resources
were found by the 1949 geological survey carried out for the
British Government by the University of Cape Town.

The first series of eruptions built up the ridge under the sea, including the base of what was to be Mauritius, and the island that
then appeared could have been as high as the present Himalayas.
Then a series of explosions partially destroyed this land build-
up, and erosion continued to reduce the ridge until a new phase
of eruptions started. The first were in the south-west of the

33

island, and, after another period of erosion, the last series of volcanic activity ensued, with a succession of thin lava flows that covered about 70 per cent of the present upland.

Deep weathering of the lavas has produced the rich agricultural soils that support the sugar cane, though the land is liberally studded with sub-angular residual boulders. Lava tunnels, some with underground streams (as at Eau Coulée near Curepipe), were then formed. The most spectacular of these is at Petite Rivière, where a passage 'the dimensions of a railway tunnel' and hung with lava stalactites twists through the rock for ½ mile.

Though Reunion still has an active volcano, activity on Mauritius ceased over 100,000 years ago. Since then submergence and emergence have taken place, so that in some places raised reefs and beaches are found as much as 6oft above sea level. Darwin was impressed with the volcanic nature of the island: 'Captain Lloyd took us to the Rivière Noire . . . that I might examine some rocks of elevated coral. We passed through pleasant gardens, and fine fields of sugar cane growing amidst huge blocks of lava.' The mountain peaks, once so mighty and still admired by most, did not impress everybody: 'Quaint and picturesque groups of toy peaks' was how Mark Twain described the worn down pinnacles.

Mauritius is set on a submarine shelf that is mainly ½–2½ miles wide, except in the north where it covers an area of 15 by 13 miles. The shelf then plunges rapidly down to the average Indian Ocean depth of 2,000 fathoms. The northern islands around Mauritius consist of stratified tuff, which was produced during the period of the volcanic explosions and disintegrated under submarine conditions.

Mauritius' interior was once a vast crater. The inside of the crater now forms the central higher plain, and the broken remains of the crater walls form mountains and ranges, all steeply inclined towards the interior, round it. The streams of lava which once ran from the central crater can still be seen,

Page 35 Gunner's Quoin, a small island off the north west of Mauritius. It is so called because its shape, seen from Mauritius, looks like the wedge used to raise a gun to the right elevation

Page 36   (*above*) Plantation House, Port Louis, photographed from behind Queen Victoria's statue in front of Government House. Opened in 1959, it is the headquarters of the Sugar Syndicate, Sugar Planters Association, and the Mauritius Chamber of Agriculture; (*right*) the bust of politician Remy Ollier in the shade of the Company Garden in Port Louis, so called because it was once the garden of the French East India Company offices

black and smoothed, on the sea's edge round the Roches Noire area in the north-east. Lesser volcanic eruptions round the coast produced other crater wall formations, such as the towering Morne Brabant promontory in the south-west and the island of Fourneaux nearby. At Baie du Cap there is a better defined formation, and the bay at Grand Port has a large crater that can sometimes be seen through the water from a boat.

FORESTS AND FLOWERS

After the volcanic activity subsided, the island became covered with dense forest, mainly of tall slow-growing trees which, apart from some birds, remained free of wild life and humans until the end of the sixteenth century. The best preserved areas of indigenous forest today lie in the south-west and in the Macha-bee forest area. Some ebonies still stand here, though in general this tree is being replaced by faster growing pines for the furni-ture industry. The latter are treated with preservatives, so that they are not weakened by insects and then destroyed by cyclones.

Nature covered Mauritius with three main types of forest: (1) the sea level palm savannah; (2) big trees such as ebony, bois d'olive (*elaeodendron orientale*), bois du natte (*mimuspos maxima*) and bois de fer (*stadtmania sideroxylon*) growing at a height of about 800ft; and (3) forests of tall straight woods like the tambala coque (*calvania major*), whose flying buttress type of roots stood up well to cyclones. Teak and colophane were also common.

When the Dutch settled in Mauritius, they ate the 'cabbage' heart of the savannah palm, dripped calou (arak) into bottles from its trunk and used its leaves to thatch their dwellings. These activities soon removed much of the coastal palm forest, and the present Conservator of Forests estimates there are now only about 100 acres left.

Among the forests the Chinese goyava grows prolifically. At Petrin near Grand Bassin, tropical heathers can be seen

C

growing on the exposed heathland. Another frequently seen tree is the travellers palm, though in the north-east forests of them have been cut down to make way for tea plantations. In addition to Round and Aux Aigrettes Islands, Mauritius has six designated national reserves—Le Puce (170 acres), Corps de Garde (22,315 acres), Le Petrin (135 acres), Sainte Marie and Montagne Cocotte (403 acres), and Bel Ombre (2,268 acres). These were established before World War II.

Though many indigenous trees and tree orchids have been lost, about 12,000 natural species remain. The best examples of the natte tree, which climbs to 125ft, can be seen in the Machabee forest area, as well as dense groves of colophane, filled in around the base of the trunk by flowering shrubs like the vieille fille, a low pink flowering bush, the ardisia crenata, a holly-like plant with white flower and red berry produced together, and the red-flowering trochetia, which is being considered as the official flower emblem of Mauritius.

The small islands to the north of Flat Island preserve fossilised remains of early indigenous forests. Dr Ayres in his early nineteenth century article 'On the Geology of Flat and Gabriel Islands' says: 'In Flat island nearly facing Round island we find fossilised remains of an extensive forest consisting of stumps of trees closely planted about two feet high, hollow in the centre to the base, and some of them two feet in diameter.'

In addition to the indigenous forests, other species of tree or shrub useful or decorative have been introduced. These include arbre corail, a tropical American shrub growing 10–20ft, with scarlet flowers; avocado; badanier, a 60–80ft high tree 3–6ft in diameter, which can be seen in the Pamplemousses Gardens and in many other places on the island; ambrivade, a 6–12ft flowering shrub with red-veined yellow flowers; almond trees from Europe; wild acacia, now growing in the Rivière Noire area; three varieties of custard apple; and xylopia richardi, a tree with small flowers and brownish pods that can be found in the woods around Grand Port. Other

38

introduced varieties include the Judas tree, jasmine, papaya, coffee shrubs in the Chamarel area, sandalwood, calabash, Mysore thorn, varieties of palm and Madagascar plum.

Flowering shrubs such as bougainvillea are widely used in gardens; and poinsettias, oleanders, hibiscus, African tulip trees, flamboyants, frangipani and cassia plants make attractive borders to many of the roads. Wild flowering creepers such as amourette, a convolvulus with longish bright red flowers, brighten up the villages, and clematis is grown widely, as is liane de Cythere, a climber with white fragrant flowers often likened to orange blossom. The heights around Curepipe are good for gardening. With the tropical climate and plenty of rain, the flower growth is luxuriant. Roses grow well, especially in the Nouvelle France higher plateau area. Strawberries, too, can be grown; small sweet strawberries grow wild in the forests, and are sometimes served in the restaurants.

In a survey of the island's grasses in 1940 Hubbard and Vaughan came to the conclusion that man had destroyed much of the original natural vegetation and little indigenous grass therefore remained. They listed 36–40 species of grasses only. The upland grassland, where rainfall is 100–175in a year, is new. Some of the lowland grasses in what was once a drier forest area are indigenous, but most are migrant. The low rainfall areas support sugar cane, which is classed as a grass, and otherwise scrub thorn and savannah. Bamboo is also counted as a grass and is much grown, particularly as hedges in the Curepipe area. It is used ornamentally in screenings and the poles are employed in creating Hindu wedding pavilions.

BIRDS

Before the arrival of man, Mauritius was an island of birds with few predators. These birds were often flightless and their extinction became inevitable with the arrival of human settlers. The most famous of the extinct birds was the dodo. The curious bird that caught the imagination of Lewis Carroll had a huge curved and gnarled beak in which it crushed the large seeds that formed its food, useless little wings and a turkey-sized body. Dodos were sent to Europe by colonists to amuse their fellow countrymen. In 1638 Peter Mundy saw two in India, and stated that they were 'big bodied as great turkeys, covered with down, having little hanging wings like short sleeves'.

Only one of these complete birds was preserved on its death, and that was a stuffed dodo in the possession of the Ashmolean Museum at Oxford. Unfortunately in a spring cleaning purge of 1755 it was accidentally thrown on a bonfire. A curator gallantly tried to save the bird but only managed to rescue the head and a leg, which are still on show.

Though extinct, the dodo can be seen in skeleton form in a number of museums all over the world. The British Museum has an almost complete skeleton, with casts of wing bones and feet; the Royal College of Surgeons has a head and sternum; the Museum of Zoology, Cambridge, has a particularly fine skeleton; the Paris Museum has an almost complete skeleton and models; Frankfurt also possesses a good skeleton; Stuttgart has perhaps the best of all, with little restoration; and Vienna's is almost complete, like that in the American Museum of Natural History in New York. The United States National Museum in Washington has a partially restored specimen, the Museum of Comparative Zoology at Cambridge, Massachusetts, has two mounted skeletons, Durban, South Africa, has a complete skeleton, and the Hachisuka collection in Tokyo has thirty specimens of bones but no skeletons.

All these, and the only known complete skeleton of a single bird in the Port Louis Museum, are the result of the work in the mid-nineteenth century of an English primary school teacher, George Clark, who worked in Mauritius. He patiently sought in the Mahebourg area, reasoning that if the bird lived on marshy ground, this was the most likely area to find traces of it. Eventually, in 1875, he found numerous bones in the

Mare des Songes, a place now covered by the runways of Plaisance airport.

The Port Louis Museum has examples of the island's other extinct birds, including the aphanapteryx or Mauritian red tail, a tall heron-like bird that could escape from its attackers more easily than the cumbersome dodo by running. It has not acquired the world fame of the dead dodo, perhaps because of the difficulty of pronouncing its name.

The dodo had a remarkable number of names—about eighty are listed. The Dutch called it the *walchvogel*, disgusting fowl, probably because it tasted unpleasant to the hungry sailors who tried it, though it could not be called a pretty bird either. Further names included bastard ostrich, monk swan, hooded turkey, bird of Nazaret and ostrich cassowary.

Other extinct birds of the Mascarene Islands include the white dodo of Reunion, of which no bones have been recovered, and similar related birds, the solitaire of Rodrigues and the solitaire of Reunion; examples of these can be seen in the Port Louis Museum. The greater flamingo (*le flammant*) is also extinct, the last one being recorded in 1880. The cattle egret (*le heron garde-boeuf*) became extinct but was reintroduced from Madagascar. Other birds have been brought in from other areas but have not lasted on Mauritius. Among these species must be numbered the Madagascar partridge (*la caille*), though a few are said to exist still on Reunion; the common pheasant; the pied crow from Africa; and the Indian house crow.

It has been estimated that around forty-two species of birds have vanished from Mauritius and its sister islands of Rodrigues and Reunion. The birds in this category include the Mauritius duck, the Mauritius darter or water turkey, the Mauritius black sparrowhawk, the flamingo, the Mauritius heron, the flightless and almost flightless herons, the flightless rails and the Mauritius parrot. The incidence of so many flightless species, as in the case of the dodo, again points to the early lack of predators on the island.

MAURITIUS

In 1912 R. Meinertzhagen published his survey *On the Birds of Mauritius*. In this he identified seventy-one species, of which twenty-eight had been introduced to the island by man, twenty-five were migrant (mainly seabirds) and eighteen were indigenous. The native species were tamer and easy to trap and kill, as had been the giant turtle and dugong, which again had disappeared from the area.

It is alarming that few indigenous birds are now left on Mauritius, and most of these are fast approaching extinction. One reason for this was the cutting down of the huge forests for boatbuilding during the French supremacy. The British, alarmed at droughts in the 1880s, bought back about 30,000 acres of forest from the concessionaires to whom the French had allocated them. With decreasing forest and damage done by monkeys to fruits and birds' eggs, many species were irreparably harmed. Of the native birds, the Mauritius kestrel (*manger de poules*) is now almost extinct, and the Mauritius fody (*l'oiseau banane*) and Mauritius cuckoo shrike (*merle*) are rare. More common are the Madagascar fody or cardinal bird, the male a flamboyant scarlet; the Manioc bird; the Mauritius parakeet, which is green with a red beak on the male and a grey beak on the female; and the Mascarene flycatcher. Flourishing species include the red-whiskered bulbul and, especially in the Morne Brabant area, the serin, a yellow sparrow-like bird.

The various types of birds introduced to the island met with varied fates. The pied crow was introduced three times from Africa and Madagascar but died out. The Indian house sparrow does well. The rose-coloured pastor died out in 1892, and the java sparrow, once very common, has now vanished. The Indian mynah (*le martin*) is extremely common on all three islands, and the waxbill (*le Bengali du pays*) thrives, along with the Kaffir golden weaver (*slug-slug*). Scarcer to spot are the grey-headed lovebird (*la petite perruche*) from Madagascar, and the Madagascar turtle dove (*le pigeon ramier*). The Indian spotted dove (*la tourtrelle*) never took in Mauritius, but in the filao (casuarina) woods along the coast the helmet guinea fowl (*pintade sauvage*) and turtle doves are found. The Chinese francolin (*la perdrise pintade*) is not plentiful, though the grey francolin (*la perdrise rouge*), originally from India, breeds abundantly. Several unsuccessful attempts were made to introduce the Madagascar partridge (*la caille*), and even the common pheasant died out. The commonest quail is the blue-breasted variety (*la caille de chine*), and ducks from Madagascar like the white-faced (*la sarcelle*) and Meller's (*le canard sauvage*) are increasing slowly. A recent introduction is the Madagascar button quail (*la caille de Madagascar*), which as yet is not plentiful.

In addition, there are a number of species which come as visitors to the island. Among these are listed the Madagascar peregrine falcon, sooty falcon, Madagascar broad-billed roller, Madagascar cuckoo, turnstone, grey plover, great sand plover, curlew, whimbrel, green sandpiper, common sandpiper, greenshawk, curlew sandpiper, sanderling, Seychelles roseate tern, Mascarene bridled tern, Indian Ocean white tern, Mascarene noddy, Seychelles white-capped noddy, Seychelles wedge-tailed shearwater, Baillon's little shearwater, giant petrel, Kerguelen whale bird, grey-backed sooty albatross and Mascarene lesser frigate bird.

The native or near-native birds of Mauritius include *la merle cuisinier, le coq des bois,* Mauritian white olive eye (*les yeux blancs* or *fit fit*), Reunion martin (*la grosse hirondelle*), Mauritian swiftlet (*la petite hirondelle*), Mauritian pink pigeon (*la colombe de mayer* or *le pigeon de marres*), Madagascar moorhen (*la poule d'eau*), Madagascar green-backed heron, red-footed booby (*le fou blanc*), Indian Ocean white-tailed tropic bird (*la paille en queue à brins blancs*), which breeds in the southern savannah area, and red-tailed tropic bird (*la paille en queue à brins rouge*), which breeds on Round Island, Gunner's Quoin and Pigeon Rock.

Pike describes the birds on Round Island:

43

At the foot of the gorge opening out to the sea, the rocks are shelving, and in the little holes in them sat numbers of Pailles-en-queues on their solitary eggs. These beautiful birds did not attempt to move away from me, but merely uttered a shrill cry and prepared for resistance if disturbed. They do not build any nest, but lay their one egg on the bare rock. It is of reddish brown, speckled with dark spots, and is about the size of a duck's egg.

ANIMALS

The only indigenous mammals on Mauritius before man arrived were several species of bat. Deer, monkey, hare and mongoose, which now make up Mauritius' wild life count, were all introduced by man. The Dutch introduced sambhur deer from Java in 1639 and these flourished so well that now big herds exist. They can be seen in the Robert Edward Hart garden in Port Louis, and excursions can be made to see herds on private estates in the Yemen area. The monkeys can be seen in the trees along the Chamarel roads, from which glimpses of deer can also be seen.

There are still a few examples of two indigenous species of the boa snake family found on Round Island. On Mauritius itself there are fifteen kinds of reptile—lizards and geckos— and 2,000 species of insects and butterflies. Nicholas Pike describes catching a river eel over 12ft long, and on his visit to Round Island:

> I captured a number of lizards, spiders, scorpions, phasmas and other insects . . . One of my comrades killed a snake of the Colubra tribe, about two feet long, and two inches in circumference. The back was mottled with black and white spots, and the belly reddish with black markings. It was what a naturalist would call an ugly customer: it does not run from you but elevates its head at your approach, and prepares to give battle. A large one was seen by one of the fishermen, who said it was six or seven feet long, and as large round as his arm.

4   THE HUMAN LABORATORY

THE animals of Mauritius are outnumbered by the humans. 'People, not minerals, are our natural sources of wealth', and 'our hidden export is cheap labour not raw materials', are phrases often used about work forces in Mauritius. The island's main problem is, in fact, chronic overpopulation. About 860,000 people live on her 400,000 acres of surface, and of these about 50,000 are at present unemployed.

Although the government has put forward sensible and purposeful plans for industrial development and diversification that could alleviate the difficulties, the unemployment problem is aggravated by the remarkably high standard of education in the island. A white collar job is everyone's dream: parents sacrifice and children study hard to gain their British academic qualifications, every library is full of youngers reading, and the competition for university places at home and abroad is fierce. But the number of jobs is not keeping abreast of the abilities available, and the discontent of qualified young men without suitable jobs has led to political unrest and criticism of the government in the MMM (Mouvement Militant Mauritian).

Overpopulation problems reflect a worthy elimination of some of the more disastrous natural methods of population control. The end of the nineteenth century records fire, cholera, malaria and cyclones decimating the population. In 1967 Mauritius was given its certificate of freedom from malaria by the World Health Organisation. With increased medical and social welfare services, the population losses have been

45

greatly reduced. Free medical treatment is given to all who go to the island's hospitals to receive it. Anaemia caused by bad diet is probably one of the most prevalent diseases at the moment.

A strenuous effort has been made to combat an excessive population rise by family planning and by emigration programmes. Australia receives about 2,000 emigrants a year from Mauritius, France still exerts a big attraction and there are 4,000 Mauritians in Natal. Since prospects in Europe and South Africa are limited, Latin America is now being studied for emigration possibilities. Family planning clinics have been set up in the villages, but make little impact on the problem.

Alongside the problem of crowding are those of a multi-racial society. For the visitor, this human laboratory of many cultures and races living closely together is a fascinating study. So far a balancing trick has been performed so that no one culture has swamped another. From this the criticism arises that too many languages and cultures are being preserved at the expense of something essentially Mauritian, to which all can contribute. The preservation of cultures was given constitutional status in the Treaty of Paris in 1814, where Britain's earlier promises of 1810, when the island capitulated, to grant freedom of culture and religion to her new French dependencies were ratified. Along with culture French law was preserved, and the Code Napoleon still exists beside a British-style legal system. In spite of about 150 years of British rule and English as the official language, French is the language of daily and more universal use, though many people are bilingual. There is always the fear that the race with the greatest numbers may sweep aside some of the older cultures in favour of its own. The Mauritius Broadcasting Company and TV helps keep alive the different languages, with times allocated to each main group. The newspapers are predominantly in French, though there are often English articles in them, and there are Muslim and Hindu papers.

46

POPULATION

In 1789 it was estimated that Mauritius had a population of 60,000, of which 10,000 were white. Though few British came to colonise after 1810, the population had reached 80,000 by 1830, still with 10,000 whites. Between 1837 and 1907 450,000 Indians emigrated to Mauritius and 160,000 returned home. Lesser numbers of Chinese arrived also in the same period.

In the last 25 years the population has almost doubled, and at the end of 1971 was 830,606. It is a young population with 57 per cent under twenty-one and 43 per cent under fifteen. The ethnic breakdown of population comprises 230,487 descendants of Europeans, Africans and Malagasies; 437,365 Hindus; 137,758 Muslims; 24,996 Chinese. There are in addition 1,000 UK nationals living in Mauritius.

PUBLIC HOLIDAYS

An indication of the attempts to keep a balance of cultures is seen in the number of public holidays held in Mauritius. Until recently there were a remarkably high number—about thirty—but now they have been cut down so that each major religious group has about two each. The list of holidays now comprises New Year's Day, Eid-el-Daka (Muslim festival, end of January), Cavadee (Tamil festival, end of January), Maha Shivaratee (Hindu festival, mid-February), Spring Festival (Chinese New Year, mid-February), Independence Day (12 March), Labour Day (1 May), All Saints Day (1 November), two Hindu festivals and the Muslim Eid-el-Fitr (November), and Christmas Day.

RELIGIONS AND THEIR CEREMONIES

The diversity of religions colours the scene with ceremonies and a variety of temples. Of the 'general population' (the

47

descendants of Europeans, Africans or Malagasies) 93 per cent are Roman Catholics, as are 44·6 per cent of the Chinese population and 4·3 per cent of the Indians. Protestants, most of whom belong to the Anglican church, are only a minority. The rest follow the Hindu, Muslim or Buddhist faiths. There is no official religion on the island. All priests are paid by the government and each church or sect receives aid from it—a factor that has divorced the churches from direct involvement in political affairs.

Hindus

Hindu shrines and temples are found all over the island. Bamboo poles carrying white or pink pennants indicate a garden shrine or temple. The Hindus set great store by their wedding ceremonies, and in many villages one sees large marquees, consisting of bamboo poles and tarpaulin, erected to hold these. One of the services the Sugar Industry Labour Welfare Fund provides is inexpensive hire of such items for weddings.

The Hindus have adopted the Grand Bassin lake as their sacred spot, the equivalent in Mauritius of the Ganges. Once a year, in mid-February, they hold the Maha Shivaratee (the Great Night of Shiva) on this spot. Processions of worshippers come from every village in the island on foot (more recently in cars and hired buses as well), usually arriving on Friday and staying overnight at the lake in large dormitories built into the hillside. They spend the night in worship and return to their villages in the morning carrying holy water from the lake for further ceremonies in local temples.

It is a slightly unnerving sight to see the massive procession of devotees, mostly in white robes, blowing crude horns and shouting religious slogans. The big feature of this festival is that many of the pilgrims carry wooden arches called 'kanvar' over their heads. These are painted white and intricately decorated with paper flowers and tinsel all a-blow and sparkling in the hot humid air, as sweat pours down the bearer's face. In 1972 cyclones battered the processions; the tinsel and gay ribbons were bedraggled, and the triumphal arches built of bamboo and palm branches over the road were buffeted, but the fervour was undampened.

In mid-March Holi is a happy and jokey celebration for Hindus when everyone squirts everybody else with coloured water and powder. For Ganga Snan in November, devotees go to the seaside to be purified by a bathe in the Indian Ocean in the absence of the Ganges. The Hindi festivals of Divali (the lights) and Dusera are celebrated in the autumn.

The characteristic Hindu religious opposition to killing animals produces scores of mangy pariah dogs that howl mournfully at night; they are survivors of new-born litters left hopefully in the cane fields by devout Hindus.

Tamils

The Tamils belong to a fiercer, seemingly masochistic, branch of the Hindu faith. In Mauritius their firewalking and sword climbing demonstrations attract big audiences; these events are announced in the local papers about 2 weeks ahead of the performances, during which time those who will be taking part in the ceremony will be purifying themselves. *Marche sur le feu—et une ascension sur 67 sabres* (firewalking and climbing on 67 swords)—the announcement is simple, the sight spectacular. Creve Coeur, a popular place for such ceremonies, lies near the foot of the Pieter Both mountain. Flags and walking crowds point the way between cane crops over fields to a clearing filled with people, and women selling food. In the centre is a metal ladder held in place by wires. A yellow-garlanded man climbs up and down it repeatedly. The rungs are frayed sharp swords, but his feet are not cut or bleeding though sweat pours off his face. Nearby is a wide pit of glowing charcoal on which men, women and children have walked without pain. They say, if you are pure enough, the feet are not harmed.

49

Mortification of the flesh in honour of their god and to prove their purity of worship culminates in the Cavadee celebration at the end of January. The *cavadee* is a wooden arch decorated with flowers and supporting at each end a pot (*somboo*) of milk. Devotees carry these, often in fulfilment of vows or to highlight their faith. They thread long skewer-like needles through the flesh of their bodies, cheeks, nose, ears and tongues for the processions through the island. A less gory festival, celebrated at the beginning of the year, is a harvest festival in which Tamils wear new clothes and offer cooked rice to the gods on banana leaves.

Muslims

Half the Muslims are concentrated in the urban areas, particularly in Port Louis where their principal temple, the Jummah mosque, is situated, and half in the countryside. Their main festival, the Muharram (known in Mauritius as Yamsey oe Ghoon), when figures and towers called *ghoons* are carried in the streets, commemorates the martyrdom of the grandson of the Prophet. Other festivals are the Eid-el-Fitr, marking the end of Ramadhan, Eid-el-Dalca and Eid-el-Adha, commemorating Abraham's offer to sacrifice his son Isaac, in which rams and goats are sacrificed and the flesh distributed to family and friends.

Chinese

The Chinese tend to live a rather closed community life. In Port Louis the main Chinese quarter is the section between Royal and Desforges streets. Chinese restaurants and clubs tend to dominate the night scene, and in general the Chinese are the shopkeepers and commercial workers of Mauritius. There are occasional dragon dances, but the main ceremony is the Spring Festival marking the beginning of the Chinese New Year. Every house is thoroughly cleaned for the day, offerings are brought to the pagodas on the Festival's eve, and firecrackers burst through the narrow streets at midnight.

The Chinese go on holiday and visit friends during this time, and are to be seen in crowds on their favourite beaches, such as Peyrébère and Flic en Flac. Although the rest of the island takes only the one-day holiday to mark the event, the Chinese tend to shut shop and visit each other, to feast over several days, though in Mauritius they do not feast for the traditional 15 days. Perhaps that is because the Mauritian Chinese have an even greater reputation for working hard that the rest of their race. For instance, in the examinations for scholarships to European universities all six places were won a couple of years ago by Chinese.

Christians

Though Christians are drawn from all ethnic groups, they mainly comprise Mauritians of French descent. The Catholic churches have a powerful section concerned with social and welfare work, and the churches are well filled with devout worshippers. The Catholics run their own schools and convents, as does the Anglican Church, which has twenty parishes on Mauritius, in charge of priests of Mauritian birth, and its own schools, though on the island of Rodrigues the two churches work together to provide educational facilities.

Although they are now separated politically, Mauritius and the Seychelles still share an Anglican church hierarchy. At the end of 1972 a new Anglican province was created—the Province of the Indian Ocean. The former Bishop of Mauritius with the Seychelles then became the first Archbishop of this province, which now covers Mauritius, the Seychelles and three dioceses in Madagascar.

Particularly venerated is the shrine of Father Jacques Desiré Laval in Sainte Croix, Port Louis. Laval came to Mauritius in 1841 especially to preach to the ex-slaves, and during his 23 years on the island is said to have converted 67,000 people.

Miracles are attributed to him and his canonisation is under papal consideration. Another place of pilgrimage is Notre Dame du Grand Pouvoir, near Mahebourg.

Christian ceremonies are now celebrated very much on European lines and can hardly be said to be as colourful as those of other Mauritian religions. A century ago, however, the celebration of Fête Dieu under the blazing tropical sun must have been an enchanting spectacle. Nicholas Pike reports:

> Files of women of every shade, from tawny to black, crowned with wreaths of roses, or white veils, or both (contrasting curiously with their dark skins) proceed leisurely up the street; delicate fair girls, dressed in the prettiest costumes, veiled, booted, all in pure white, but in a shower of ribbons and flowers that flutter down from the silken embroidered banners they bear.
>
> Very small fairies, aptly termed 'Les Anges' trip along carrying baskets of flowers, and they also wear dainty white satin shoes. I was told that only a few years ago a number of little children, chosen from the best families, were always present dressed in a white gauzy texture with wings and their pretty little feet bare. Heat and fatigue and often a heavy shower wetting them through caused such severe illnesses that generally one or more fell victims to the cruel practice, so it has happily been abandoned.

Creoles

Originally the word 'creole' meant anyone of European descent born outside the mother country of France. In Mauritius it has come to mean anyone of the original inhabitants apart from British, Indians or Chinese. The creoles are of African or Malagasy descent and their African origins are expressed in a prevalence of black magic in the villages and the hip-wriggling sega dance. The sega (pronounced saygah) is performed for tourists at the hotels, or on the quayside when a cruise liner arrives, but this lacks the earthy ecstasy of the natural sega, which develops, as the rum goes round, into an orgy. For this

Page 53 *(above)* The Jummah mosque in Port Louis is the centre of Mauritius' Moslem worship. Its carved doors are inlaid with ivory; *(below)* the Chinese are an important community and own many shops. This is the pagoda in Port Louis

Page 54 (*above*) The theatre at Port Louis built in 1822. Intended for opera, it is mostly used for amateur drama performances today; (*below*) Government House, the ground floor of which was built by Mahé de Labourdonnais in 1738. The present building dates from 1771 and is used for cabinet and government meetings

reason it is mainly performed on Saturday nights or on the eve of public holidays. The Sunday hangover is a fact of village life, and the fishermen of the coastal villages never do much work on a Monday.

It is possible to see a sega session if you approach the headman of the village and offer to pay for the wine and rum needed to get the party going. With a rising incidence of alcoholism both the authorities and the local Roman Catholic priests are naturally averse to the dances, which undermine their Sunday morning service attendances. Enough drink for a sega works out at several dozen bottles of wine and rum—probably the illegal brew. The original instruments, now used less and less, were tin cooking pots and spoons, helped out by hand clapping. The simplest rum-sent segas have this background; the dancers moving to the sound of the tintinnabulations of spoons, and seeds shaken in a tin. The pace starts slowly and works its way up to a frenetic beat. Other instruments used are an iron triangle, and a guitar—once made from strings attached to a calabash gourd but now more likely to be an Hawaiian or an electric guitar. Words in creole accompany the rhythm, which starts with a stately drum-like summons to the villagers, who come and congregate round the singing and drumming. New musicians take over when the first group becomes exhausted.

VILLAGE LIFE

Apart from Port Louis and the towns described in Chapter 10, there are ninety-eight villages on the island. These are grouped under three area councils for welfare and local administration; and each village has its own council and chairman, most members being elected, and the rest nominated by the government.

Life and entertainment are simple. Most, but not all, villages have electricity. Many have a village hall or *baitka*, or a small meeting place apart from the family planning clinic or dis-

D

pensary. On the sugar estates the hall and recreational amenities are provided by the Sugar Industry Labour Welfare Fund. There is often a cyclone shelter as well. In the village hall a TV set is the main attraction, and the men also gather here to play dominoes, table tennis and talk. During the week the hall may serve as a centre to teach young girls how to sew. Most big villages will have a football pitch.

The houses, thatched with palm or roofed with tin, each have little gardens in which fruit and vegetables are grown and chickens kept. Outside many houses is a large smooth stone, the washing stone, on which clothes are rubbed clean, though mostly washing will be done in the little streams and rivers around the island and clothes spiked to dry on nearby bushes. Water usually comes from a communal tap. Life revolves round the stores, mostly Chinese-owned, where a few grains of rice and a screw of spices may be bought. At the Tabagie, a tobacconist-cum-bar by definition but now just a tobacco shop, one or two cigarettes may be bought at a time. On Sundays the men lounge in the shadow of the banyan trees, go to the bars for the rough white local rum or play games in the village hall, while the children of ambitious parents, particularly the girls, march past with school books on their way to extra private lessons.

If he can, each villager keeps chickens, and perhaps a pig and a cow. The women rise early to collect fodder for the cows from the edges of the cane fields. It is said that some 30,000 cows are kept this way.

The fishing villages are built around the sheltered bays, and boats drawn up in back gardens or left anchored within the reef. The fishermen's pirogues are long wooden boats resembling dugout canoes with both ends pointed. Some are powered by outboard motors, others have simple sails and glide within the reef looking for fish. As a sideline, fishermen now collect shells for the tourists, going out early to dive on the reefs. They set up small stalls in front of their houses along the roads, and their wives usually organise sales, pricing the rarer shells by an international shell book.

FAMILY LIFE

The villages tend to be intensively communal, and the better-off try to escape from their restrictions. But it is difficult. 'No one is independent in Mauritius, every one has a society', said a businessman, and this is true. There are few secrets kept on the island. 'If you dial a wrong number in Mauritius, you can even so talk for half an hour', sums up the forced gregariousness of life there. The people hide themselves in their houses behind thick bamboo hedges in Curepipe without street numbers or road signs. At weekends they drive to their *campements* or beach-side bungalows, secluded among the casuarinas along the coast. It is still rare to entertain in a restaurant. With plenty of cooks and servants, entertaining at home is the custom, and the islanders, rich and poor alike, are generous in their hospitality.

The family unit is sacred and the strong family ties make Mauritian society seem somewhat old-fashioned. Girls are sheltered and marriages are arranged between families where possible. European, Chinese and Indian children alike respect and obey their parents, who are in return devoted to their children's advancement, the poorer families scraping to send a child to school and perhaps university. This has made the youth in Mauritius more receptive, less restless and revolutionary than that in other countries. But there are too many youngsters educated above the jobs they can find. Every year at least 10,000 new jobs are needed, a factor noted in the government's 4 year plan.

Those who win the coveted prizes to European universities in Britain, France and Russia are exposed to much freer and wider societies, and on their return to Mauritius find the narrowness irksome, though they are usually quickly absorbed

57

into the family circle again and smartly married off. Their restlessness, however, is expressing itself more and more in political unrest. Many girls go to work as nurses in Britain and a scheme is being promoted to teach them more about their country so they can act as 'ambassadors' for Mauritius abroad.

5 ADMINISTRATION, LAW AND EDUCATION

CONSTITUTION AND POLITICS

IN miniature the constitution is similar to the British system. Universal adult suffrage was introduced in 1958, and the present constitution came into force on 12 March 1968 with independence. The parliament consists of seventy members, including the Speaker. Elections should by law be held every 5 years, though the next election has been put back to 1976. Apart from the Prime Minister, the constitution lays down that there are not to be more than twenty ministers. In the Assembly seats represent the constituencies of Mauritius, each of which returns three members (the best way of ensuring ethnic representation), and two members for the island of Rodrigues. The Prime Minister governs with reference to the Governor General on occasion. The latter acts as the representative of the Queen, since Mauritius is an independent governing member of the Commonwealth.

Until 1955 there was in effect only one political party—the Mauritius Labour Party, founded in 1936 by a creole. The head of this party became the present Prime Minister, Sir Seewoosagur Ramgoolam, and his followers were in the main Indian. In 1955 the PMSD party (Partie Mauricien Social Démocrate) was set up, and it was headed in 1966 by Gaetan Duval, a young and brilliant lawyer who is the present foreign minister. The PMSD, representing the white and creole communities, opposed independence in 1968, fearing domination by the Indian population and preferring an association similar to that held now by the Seychelles. In November 1969, however, the two parties joined to form the coalition government that is still in power.

Five other political parties exist: the CAM (Comité d'Action Musulman), founded in 1958 and the official Muslim party, some of whose members joined the PMSD; the Independent Forward Bloc, also set up in 1958, which is preoccupied with constitutional reform, including the creation of an Ombudsman, co-operative economic measures and social security; the Tamil United Party; the UDM party (Union Démocratique Mauricienne), set up by dissidents from the PMSD grouping at the end of 1970 and consisting of a dozen members under Maurice Lesage, who is the official 'leader of the opposition'; and the MMM (Mouvement Militant Mauricien), which made a dramatic impact in 1970 by gaining important seats under its creole leader, Paul Berenger. The last-named party is left wing and revolutionary, inciting dock strikes and attracting much discontented youth to its policies. During the existing state of emergency the MMM has been forced underground and its newspaper closed down; some of its leaders are watched by the police and others are in prison.

HOSPITALS AND SOCIAL WELFARE

The island has eight general hospitals with a total capacity of 2,778 beds, and includes one of the most modern in the Indian Ocean—named after the Prime Minister, Sir Seewoosagur Ramgoolam, and set in wooded country near Pamplemousses. Though herbal medicine and illegal nursing are practised in the villages, medicine is free for everyone provided they go to the hospitals to receive it. There is a psychiatric and a leper hospital, and twenty-four hospitals run by the sugar estates. The hospitals are too few to cope. There are few private clinics, and those that exist are often too expensive for most people.

Villages have family planning centres, which, in addition to the strenuous programme for birth control, give post- and ante-natal care, and occasionally there is also a dispensary for minor medical care. Fifty-three dispensaries and mobile centres exist in the island.

The sugar estates are responsible for the medical care of their 70,000 workers, and they have set up their own welfare centres (see p 146) under SILWF (Sugar Industry Labour Welfare Fund).

POLICE AND CRIME

The confines of a narrow traditional society make behaviour most circumspect. Visiting sailors find consolation in the small hotels and houses along the Petite Rivière coast or in certain parts of Port Louis. Though kept underground, the incidence of immorality has alarmed some Mauritians, who are forming a morality committee.

In a densely populated area like Mauritius, crime is more concerned with possessions than passions. The island has a British-style police force of about 1,500 men, plus a Police Riot Unit and a Special Mobile Force (see p 134). There is a prison near Quatre Bornes and the death sentence for murder is still law.

Theft is the commonest crime, whether of fodder from the fields to give the cows, or of money and possessions left carelessly around. Tourists are warned in a special leaflet issued by the Tourist Board to take care of their valuables and lock their cars and hotel rooms. In addition to crime prevention, police stations act as information centres to give directions in the unnamed streets, and in the villages police lecture to children on road safety and courtesy to visitors. Under the state of emergency, which has existed since the riots and strikes of December 1971, some of the more extremist politicians are kept under police watch or imprisoned.

MAURITIUS

The law is Franco-British, a blend of the Code Napoleon and the English common law. It is further complicated by diverse standards of measurement.

Three judges, a chief justice and senior puisne judge administer the law in the Supreme Court, which also houses courts of civil and criminal appeal. It is situated near the City Hall in Port Louis, and in front of it are the Port Louis District Courts. The Supreme Court is also the seat of the Assizes, which meet four times a year. All the island's solicitors and barristers have their offices in Port Louis around the court area. Many of Mauritius' leading politicians are barristers who read for the law in London.

The University of Mauritius, founded in Le Reduit with the help of a £3,000,000 grant from the British government in 1965 and a gift of £3,000 worth of books, is tackling the problem of practical education rather than intellectual training. The three faculties cover agriculture, administration and industrial technology.

The School of Agriculture is important in relation to Mauritius' dependence on sugar for its economy; it is linked with the Faculty of Agricultural Science at Reading University, England. The School of Industrial Technology was set up with the aim of fulfilling the demand of industry for trained manpower, and offers a variety of courses in engineering and technological subjects, many of the students being sponsored by local industries. The School of Administration covers public administration, local government, and professional and economic studies. It has connections with Manchester and Birmingham Universities in England, and associated with it is a Centre d'Administration Entreprises run in co-operation with the University of Aix-Marseilles. A feature of the university is the large number of short courses and seminars that offer part-time training for industry and agriculture, in order to produce additional trained manpower for the local economy.

There are 500 full-time students, and 1,200 part-time students who come to special evening classes. Courses are held in such practical work as sewage construction (a seemingly endless procedure in the narrow roads of Curepipe) and motor repair work, so that maintenance can be carried out on the island. Evening courses for bank clerks are also held.

Plans are being made to extend the university faculties. An Institute of Education is planned for 1973, together with a school of bilingual studies, as well as an auditorium financed by the French government. Attached to the university, but physically separate at Rose Hill, will be a Hotel School at which UN experts will train staff for the new hotels the island is planning. It is hoped that the Hotel School will eventually be able to act as a small hotel, whose paying guests may give students practical experience, and at the same time reduce its running costs.

SCHOOLS

A visiting educational expert when asked what impressed him most about education in Mauritius remarked: 'There's an awful lot of it.' Nearly 14 per cent of the Mauritian budget is spent on education, which is based on the British system, though many of the classes are carried out in French.

In 1815 the Independent Protestant Rev Père Jean Le Brun founded the first free schools and chapels for the workers of Port Louis. Now primary education is free and available to all, though not compulsory. However, 86 per cent of children receive primary education. Primary schools are overcrowded and conditions are aggravated when there are severe cyclones, as in 1960 and 1962, when buildings were damaged and a double shift system had to be worked. In 1971 there were 347 primary schools teaching 148,682 pupils, with a ratio of 32 pupils for each teacher. The education authorities are divided into four, representing Roman Catholics, Church of England, Hindus and Moslems, but the teachers are paid by the Ministry of Education. Around 4,000 primary school children attend private schools that do not receive government aid.

Primary education lasts 6 years, from the age of five, but children can compete till the age of twelve for scholarships to the secondary schools. The teaching of arts and crafts forms an important part of the work done at primary schools. Secondary education is of grammar-school type leading towards university. A number of parents feel their children are in need of further tuition and send them to private lessons, many of which are given by the full-time teachers. Since 1951 a number of the private secondary schools have been judged 'approved secondary schools' and been aided by the government. In 1964, for example, there were four government secondary schools, thirteen subsidised schools and ninety-four private unaided schools. The government maintains a school health service with medical and dental treatment, which is free in the government primary schools, as is milk. By 1971 there were 137 secondary schools with a total of 46,897 pupils, of which 36,845 were non-aided. At this level, the pupil/teacher ratio is 27 : 1.

PORT LOUIS

IN the days before sophisticated navigational aids, sailors making for Port Louis harbour would take a fix some 30 miles out on the rock that resembles a ball balanced on the nose of the Pieter Both mountain. When it appeared dead centre in 'the window', a square gap in the mountain behind the town, they were on course to slip in through the reef. It was a relief to see these marks, for Port Louis, before the Suez canal was opened in 1869, was a welcome resting place on the long voyage to India or Australia. Once the Canal opened (ironically Madame de Lesseps was Mauritian and Mauritians worked on the canal), the importance of Port Louis declined, though for a while it was a useful coaling station.

Port Louis is the capital and only city of Mauritius, set between the sea and the semicircular Moka range. Today the expanding city of 140,000 inhabitants covers 16sq miles and has spread outside the protective arms of Citadel Hill and Signal Mountain, which confined it originally. It became a municipality in 1850, and was granted city status by Queen Elizabeth II in 1966. The residential areas of Curepipe, Moka, Rose Hill, Beau Bassin, Quatre Bornes and Vacoas, some 20–30min away by car, are joined to it by the island's sole motorway and wide roads. Mahebourg, the main town (then called Grand Port) when the French arrived, is on the south coast and the only community in that area qualifying for town title, though it is now crumbling and neglected.

Mahebourg was named after Bertrand François Mahé de Labourdonnais, the French governor, revered as the man who really created the colony of Mauritius. In 1731 his predecessor, Nicolas de Maupin (1729–35), chose Port Louis as more cyclone-proof than Grand Port, and Labourdonnais turned to building up its strength at the beginning of his governership in 1735. It had already been named after Louis XIV. Labourdonnais realised that the south-west winds made Mahebourg a less safe haven than the deeper mountain-backed harbour of Port Louis.

Today Labourdonnais' statue stands at the inner jetty of the harbour, backed by an avenue of palms and facing out to sea; and in front of him sugar ships, Japanese tuna trawlers, Russian fishing boats and visiting cruise liners ride at anchor. Port Louis still looks to her harbour for news and excitement.

Even on days when no new ship enters, the crowd circles round the dock area, shifting and lively. Boys wander among the cars selling sweets and postcards from trays slung round their necks; shoe cleaners, always busy, bend their heads to the ground; Chinese trot by with baskets slung over their shoulders; Indians push barrows of goods to market; loafers stand in the shade of the corrugated roofs that edge the buildings of the shipping agents; street traders hover hopefully in door-ways on the lookout for passing tourists; and women in saris or bouncy mini-skirts try to avoid the deep gutters awash with garbage after a tropical storm.

GOVERNMENT HOUSE

Behind Labourdonnais' statue, the wide main avenue of Place d'Armes, stately with palms and plants, leads direct to Government House. This was the governors' residence from the days of Labourdonnais, who built the ground floor in 1738, until Le Reduit was built in 1749, when the governors moved to that residence's cool heights and gardens from the heat and humidity of Port Louis. The first telephone line in Mauritius connected

Government House to Le Reduit in 1883. Government House still remains the seat of government. A new parliament chamber has been tacked on to the back of the old French colonial building of wide balconies, three-sided round a courtyard and the oldest intact building in Mauritius.

The courtyard contains a statue of Sir William Stevenson (1857–63), a British governor who claimed for Mauritian officials equal privileges with British officials, and in front of him, white, and happily unblessed by pigeons, is a statue of Queen Victoria whose plaque records how deeply her subjects in Mauritius mourned her passing. Facing Government House are the palm-shielded statues of Governor Sir John Pope Hennessy, who became popular locally with his policy of 'Mauritius for the Mauritians', and Sir William Newton, a Mauritian lawyer who defended Pope Hennessy before the Privy Council when his policies were under scrutiny.

The rooms in the old part of Government House are filled with portraits of British kings and queens, and the darkly panelled dining room of the members of parliament is still fitted with Victorian glass lamps. In the days when it was the governor's office there was no dining room, and luncheons were sent across on a tray from La Flore Mauricienne, a restaurant that is still in existence. The throne room, which runs across the centre between the two wings, is now the banqueting room. The cabinet still meets in old Government House. Leading directly from the old house is a sharply modern building, grey stone and rounded, starkly at variance with the paint-peeled balconied two-storey buildings that make up most of central Port Louis.

LEGISLATIVE ASSEMBLY CHAMBER

This round building, built in 1965, is the parliamentary assembly building. It reflects the British parliamentary system in miniature. Government ministers and opposition sit in semicircular ranks on scarlet chairs in a wood-panelled room fitted with blue-grey carpets. Distinguished visitors' and public galleries are set at the back. The official language is English, though some opposition speeches are made in French, possibly just as a way of opposing. The offices of the Prime Minister and other senior ministers are situated in the parliamentary assembly building also.

CITY HALL

Port Louis' other modern building, also possessing a circular council chamber, is the City Hall. This was built in 1962 and consists of a main four-storey block, council chamber and single-storey flanking office buildings. The ground floor contains the city library, and above it are the mayoral offices and reception rooms. To the right of the entrance is a small city information office.

The city is run by a council of sixteen elected members, mayor and deputy mayor. They meet in a wood-panelled circular room furnished with grey carpet and dark blue chairs (each carved with the coat of arms of Mauritius). On one wall hangs a large wooden coat of arms of the city beautifully carved by a one-armed Indian craftsman of the island: over two sea horses there are clasped arms, the original symbol of the city before the present coat of arms was granted in 1961. The city motto is *Concordia et Progressio*. The famous letter from Queen Alexandra to the Mayor of Port Louis, addressed to 'Mauritius, West Indies', has been framed and is hung in the City Hall.

Outside the new City Hall stands a stylised modern concrete clock tower, which has a twisting staircase to symbolise the days when Port Louis' wooden houses were often in danger from fire and a watchman would be posted atop a tower. Port Louis has always been subject to disaster—to cyclone (1892), fire (1816 and 1893) and malaria (1867).

The Line barracks in the western part of the city, built by the French in 1736 as barracks for 6,000 men, still impress with

68

the length and strength of their walls, which enclose the area now used as the police headquarters. Between the Place d'Armes, where once the soldiers paraded, and the Line barracks runs La Chaussée, an old street but now lined on one side with modern office blocks that house the British High Commission and Mauritius Tourist Board and Information Office.

These offices look out on a charming garden of palms, creeper-clad banyan trees and a collection of busts and statues of prominent citizens sculpted in the French manner. These include Remy Ollier (journalist), Adrien D'Epinay (politician), Leoville l'Homme (poet and critic) and Brown-Sequard (scientist). This is the Company Garden, so called because once it was the garden used by employees of the French East India Company, which controlled the island until the French government took over in 1767. The garden saw service as a market and fairground before settling down to become a haven of cool stone seats and shade in a hot city. Among the tall trees is a small rustic-style building that houses the Mauritius Boutique, where handicrafts from all over the island are displayed and sold.

MAURITIUS MUSEUM

On the other side of the Company Garden is a small building housing the Museum, Mauritius Institute and Library. Dating from 1884, the museum contains a remarkably impressive array of shells and seabirds, indigenous and exotic flora and fauna of the Indian Ocean. It acts as a fine introduction to exploration of the island and its varied creatures and plants. The shell collection is particularly striking, though some of the rarer shells, such as a cone worth 8,000 rupees, are copies, the originals being locked up in the director's office. A stamp exhibition, including reproductions of the famous Mauritius penny red and twopenny blue, has recently been added to the collection. The pièce de resistance is, of course, the dodo, though the museum also has another extinct bird, the solitaire type of dodo, this time from Rodrigues, and the extinct aphanapteryx (the Mauritian red tail). There is a huge tortoise from the Seychelles over 200 years old, who came in 1766 to Mauritius and died in 1918 'of an accidental fall'. On a smaller scale are the collections of forty different types of sea urchin found in the island, as well as the many different starfish and butterflies.

Above the museum, reached by an outside staircase, are the library and Institute meeting rooms, lined with portraits of the founders, men like Charles Telfair who did much to modernise sugar production, Brown-Sequard, the American scientist who lived in Mauritius, and Pierre Poivre, who did much to originate the island's intense interest in its own natural history. The fine collection of works on Mauritius is worth studying and the required permission of the librarian is readily obtainable. This library, like the city library, is full of brown-skinned boys eagerly bent over their books.

GOVERNMENT ARCHIVES

Anyone with Mauritian ancestry or an interest in the minutiae of its social history can explore the remarkably complete collection of archives in the Archives Office, located in Sir William Newton Street near the City Hall. Under the archivist Auguste Toussaint, the Mauritian historian, and his successor Dr Adolphe, worldwide research was carried out to establish the whereabouts of records and historical data on Mauritius. The results of 10 years' labour are collected together in the *Bibliography of Mauritius 1502–1954*. The fire- and cyclone-proof archives contain notary records going right back to the first years of the French settlement. The archives also keep issues of the Mauritius newspapers and have a copyright library. Old maps of the island, a letter from Darwin, seals of various French and British kings and the original coat of arms granted by Edward VII in 1906 can also be seen.

Page 71 Curepipe: (*above*) the Park Hotel, in the style of an old French colonial house, was the first opened on the island and has 80 rooms set in 6 acres of gardens; (*below*) the Town Hall, another traditional French colonial style building. In the right foreground is a statue of the lovers Paul and Virginie showing Paul lifting Virginie over a stream

Page 72 A palm alley in the Pamplemousses Gardens. Fifty-seven acres of gardens, set on the former site of the country residence of French governors, they were later amongst the most important botanical gardens of the world

HEALTH IN THE CITY

In 1828 Rosalie Hare, the young wife of the captain of the *Caroline*, arrived in Port Louis on her way from Tasmania to London. In her diary of her 18 month journey the girl reported how gay Port Louis was, how everyone loved music. Bands performed in the Champs de Mars. Playing the harp and guitar was popular in private houses, and a constant flow of visitors from ships enlivened the social scene.

Alas, Port Louis has changed! In the great malaria epidemic of 1867, 200 a day died in the city, and in all a quarter of the population died. Fear of the disease, which seemed worse at sea level, drove the population to the cooler healthier hills behind the city. The pianos ceased playing in the elegant verandahed houses with their beautiful high walled gardens. The 1,300 carriages paying tax in the 1830s moved away, and Port Louis became a city for work, not middle- and upper-class residence.

Today the men are detergent-bright advertisements in clean white cotton shirts and carefully pressed trousers. Young executives wear light correctly tailored suits as they hurry knowingly along, for no one ever seems to stroll in Port Louis. Indian women (even in the fields) wear bright new saris— those bought for a wedding last week will be worn for work this week, and not put aside for 'best'. Chinese and French girls wear the latest European fashions.

Mauritius received its world health clearance from malaria 100 years after the great epidemic, and people do now live in Port Louis—Chinese in the Royal Street area and Muslims in the Plaines Vertes area—but it is still the snob thing to live up in the hills behind. More people, however, are settling along the nearer coastal areas, such as Pointe aux Sables, which is studded with dubious hotels and houses catering for sailors.

The working day in Port Louis is remarkably short. It starts

E

73

officially around 9 am, breaks off for a brief lunch hour from
12 to 1 pm and then ends in a homeward-bound traffic jam
around 3.30 to 4 pm. By 4.30 the streets are still—a good time
to stroll round and look more closely at the buildings.

THE THEATRE

It is a charming pillared building almost opposite the City Hall
and behind the Government House complex. Set back from the
road, it boasts a huge Paris-style street lamp, all elaborate
curled ironwork and globular lights. Though the lamp is of
the same period as the theatre, local legend says it was brought
along with a guillotine to Mauritius—symbols of the French
Revolution on which to hang or decapitate the aristocrats. But
the only Mauritian victim of the Revolution was a goat on
which the guillotine was tried out at a lively meeting.

The theatre was built to French designs, but its foundation
stone was laid by the first British governor, Sir Robert Farquhar,
and it was opened on 12 June 1822. Originally it seated 850
people (now about 450, as the top tiers are not used) and is in
classical style, with boxes for the governor general, mayor and
councillors. A big chandelier hangs from the umbrella ceiling,
which is decorated with cherubs and surrounded by the names
of European opera composers. In 1852 the theatre became
municipal property. The French companies that once came
frequently now come rarely, and the theatre is used for local
amateur productions. A special stamp was produced to mark
its 150th anniversary celebrations in 1972.

CHAMP DE MARS

The Champs de Mars was once the main meeting place on the
island. People strolled in its formal gardens, bands played and
soldiers paraded. Situated on the north-east edge of the city, it
is almost surrounded by hills. In 1812 its parade ground was

given to the Mauritius Turf Club to form a race track. The club was founded by Colonel Edward Draper, and continues to offer a popular series of meetings, mostly with Australian jockeys riding from May to October. But families continue to walk in the Champs, where there is a children's playground. At weekends the place is full, its occupants for once walking slowly and not at the customary Port Louis pace.

In the middle stands Edward VII's statue, sculpted by the Mauritian Prosper d'Epinay, a friend of the king. At the north end stands the obelisk tomb of Count Malartic (1792–1800), a popular French governor, whose memorial was completed by funds raised by Lady Gomm, the British governor's wife, who was renowned for using the now so valuable Mauritius 'Post Office' engraved stamps on her invitations. The obelisk was blown down by the 1892 cyclone, which destroyed one-third of Port Louis.

THE CITADEL

One of the best views of the Champs de Mars may be had from the Citadel on the hill immediately above, which can be walked or driven over on a rough road. The Citadel or Fort Adelaide was built by the British in 1834 as a barracks in case the recently conquered French inhabitants revolted over the abolition of slavery. Its solid grey stone walls and two-tier rooms are reminders of the British conquest. Its walls give excellent views down to and over the city and across to Signal Hill, which once had a signal station and watch post for arriving ships. The Citadel is being developed (as a tourist viewpoint), with a new road to its summit.

STREETS AND SIGNS

With few modern high-rise blocks and a love of flowers and trees, Port Louis retains a gracious old-fashioned look in spite of a pressing need for fresh paint and general refurbishing. Its streets are built on a grid pattern and most have street names, remarkable for Mauritius. As everywhere in Mauritius, the lack of signs is so well known that people are happy to walk with strangers to their destination in a most charming and friendly way. In Port Louis young workers will leave their desk or shop and guide you along several streets. In Curepipe, where road signs disappear into bamboo hedges, one stops at the corner Chinese shop to enquire, and the proprietor will know the name of everyone in the street. The names of the streets reflect Mauritius' mixed ancestory and racial hotchpotch. They commemorate favourite historical figures: Farquhar, Labourdonnais, Pope Hennessy, Sir William Newton, Bourbon and Pasteur all turn up frequently in many areas of the island's communities.

MULTI-RELIGIOUS MONUMENTS

The sightseer can see Chinese pagodas—one on the hill near the Reine de la Paix monument to the Virgin Mary for peace during World War II, another to the south-west near the Robert Edward Hart gardens, where there is a cage of Sambhur stag and deer. The Jummah Mosque in Royal Road, begun in 1852, is the pride of the Muslim community. The Catholic cathedral of St Louis stands near the Law Courts, and outside it is the statue to St Louis, the city's patron saint, erected in 1896. Nearby are the remains of a fountain dating from 1788. Inside the cathedral is the tomb of Labourdonnais' wife and daughter. The city celebrates St Louis' day on 25 August.

The Anglican cathedral of St James in Poudrière Street is set in the building once used as a powder magazine. The walls of the nave are so thick—almost 9ft—that it was possible to place full length beds in their window sills during bad cyclones. In the nineteenth century the church was often used as a cyclone refuge, though now special shelters have been built all over the island.

SHOPPING

Like the churches, the shops reflect the different races. There are smart shops for duty-free cameras, Swiss watches and Mauritian-made jewellery, such as that of the Poncini Brothers, whose father came from Switzerland over 40 years ago. Indian traders round the docks sell saris, ready-made clothes, silver and gold jewellery, batik work and sandals. In the Chinese section between Royal and Desforges Streets, jade, porcelain, ivory and Chinese foods are crammed into small shops without thought of temptation by display. Wares, however, are cheaper there than up at Curepipe or Rose Hill, where the tourists go to shop.

MARKETS AND NIGHT LIFE

All races mix in the markets. There is a small market on the road south out of town, but the Central Market is situated near the dock entrance between Farquhar and Queen Streets. Except for weekend afternoons it seethes with buyers and sellers. In early morning the main central paved way between the two covered sections seems carpeted with fresh vegetables and fruit— small sweet Victoria pineapples, mangoes in season, wood strawberries heaped in plaited baskets, onions, and pommes d'amour, the small cherry tomatoes popular on the island. Sari-clad women crouch by the artistically piled arrangements. Under cover is the meat and fish market, and on the other side the pungent crushed spices for curries marry with the stable-like smell of the baskets and *tentes* (little plaited baskets made from vacaos leaves) in which many workers carry their lunch and which visitors use to collect shells when beachcombing. Little bundles of vegetables ready selected to make a soup are widely sold, as are the *pains de maison*, big crusty rolls.

One corner houses the herbal cure men, their stalls stacked with fresh green herbs carefully labelled with names of the diseases they should cure. One stall, a father and son team, boasts 50 years of experience, while their neighbour has for 43 years been selling tisanes to foreigners, who have all been '*très satisfait*'. Tisanes cure a sensational range of illnesses according to the proffered cards, including dandruff, gout, ulcers, stone, gas, eczema, trances, tension, hernias, and '*d'autres cas particuliers à des prix raisonnables*' (other individual cases at reasonable prices).

The Port Louis market's reputation is still such that men are sent there shopping by their wives, if they work in town. In 1867 Charles John Boyle published his book *Far Away, Sketches of Scenery and Society in Mauritius*. Boyle, the chief commissioner of the railways and in charge of constructing the system for the sugar estates, lived at Moka above the town, and complained of being constantly pestered with shopping lists for his wife and of spending his carriage trips home with parcels of food piled about his feet.

The combination of office workers buying their modest lunch at the market and the early mass afternoon exodus has not encouraged the creation of good restaurants and night life. At night the Chinese quarter is the liveliest, with two casinos—a new one behind the well known Lai Min restaurant in Royal Street and, more characterful and noisy, the Port Louis Amicale club across the street, where in grubby surroundings Chinese and visitors play quatre quatre (fantan) rapidly and noisily with deep concentration hour after hour, while a waiter circulates with drinks. On the balcony of Lai Min one can sit and watch the oriental world go by, but on the whole Port Louis lacks these 'somewhere to sit' places the visitor often craves. In an island addicted to gossip and endless rhetorical discussion there are no cafés in the capital. A few small restaurants cope with businessmen's lunches. The best known and oldest established is La Flore Mauricienne (named after a locally famed ship), now in the Anglo-Mauritius Insurance building, Port Louis' tallest building. On the top of this is the City Club, where businessmen can take friends for drinks and a simple self-service lunch, and enjoy one of the best all-round views over the city.

One haven for the visiting seaman in Port Louis is the Merchant Navy Club in Rivière Street. This retains the charming secluded gardens of the Port Louis houses before the malaria exodus, and is surprisingly elegant and spacious for a building so near the docks.

THE DOCKS

The Port Louis docks, started by Labourdonnais in the Trou Fanfaron, where boatbuilding is still carried on, were expanded by the creation in 1860 of the Mauritius dock, restyled the New Mauritius dock in 1876, and the Albion dock in 1870. Eleven big ships can anchor in the main harbour, which has few dredging or shifting problems, and two ships can berth alongside. One of these berths is occupied every month by the MV *Mauritius*, a 124 passenger ship with twenty-four cabins which links Port Louis, Rodrigues and the Seychelles.

The natural harbour gives good protection, though ships stand out beyond the reef when cyclones threaten. With the growing size of cargo ships, the harbour will be deepened. To the north of the harbour an area known as the Mer Rouge is being filled in to create 250 acres of land for new petroleum stores and industrial development. At the tip of this is Fort George, one of two fortresses once guarding the harbour entrance. Cannon on the Ile aux Tonneliers on the right of the entrance once protected the harbour, and a thick chain was stretched from that island across to Fort Blanc on the left to prevent ships entering. There are plans to build giant granaries and bulk sugar stores round the harbour. At present sugar is exported in bags, the rate of loading being about 1,000 tons a day. The docker starts about 7 am, takes an early break from 9 to 10 am and finishes about 3 pm. Seventy-five ships, mostly

Greek, are needed between July and February to take away the sugar crop. Near the old ship agents Blyth Brothers, whose solid premises have stood on the quayside for 100 years, are piled long cane baskets in which pigs are brought alive from Rodrigues to help the island's meat supply.

7   INTELLECTUAL LIFE AND
         ENTERTAINMENT

PRESS, RADIO AND TV

'THE second industry in Mauritius is gossiping' is a common comment. Introverted discussion of local affairs and history is a national pastime, reflected in all the daily papers the island produces. They cost little and there are numbers of them, though the circulation of each is small. The doyen is *Le Cerneen*, founded in 1832; it is popular with Europeans and creoles and expresses the views of the PMSD (Partie Mauricien Social Démocrate). Other important papers with larger than average circulations are *Le Mauricien*, founded in 1908, which calls itself 'independent'; *Advance*, a left wing paper; and *L'Express*, more a supporter of the Government. Among the other of the dozen daily papers are *The Star* and *L'Aube*. About twenty periodicals of various kinds are produced in the island in six different languages.

Before World War II broadcasting in Mauritius was under the control of a radio station—Radio Maurice. In 1941 two more stations were created—Le Poste Radiophonique d'Ile Maurice, an experimental station, and Radio Liberation. In 1944 all stations were taken over by the government, which founded the present Mauritius Broadcasting Service, the responsibility for which was transferred in 1962 to the Ministry of Information, Posts, Telegraphs and Telecommunications. Broadcasting time amounts to 116 hours a week from three 10kW transmitters to 83,778 licenced sets. There are five 15kW

81

transmitters for the TV service, which is run by the same organisation from the same studios in Curepipe. There are now 57¾ hours of transmission per week and the island has 2,685 licenced TV sets.

The programmes are divided into periods of English, French, Hindustani and Chinese language. In addition broadcasts are also given in Urdu, Tamil, Telegu, Gujarati, Marathi and a couple of Chinese dialects. The programme content is mainly recorded music and transcriptions from the British and French broadcasting systems. Local live programmes, discussions and news are increasing. The daily schools service relies much on the BBC overseas schools service for material.

INTELLECTUAL LIFE

The production of books about Mauritius is staggering, though most of these are in French. The national archives in Port Louis have produced a *Bibliography of Mauritius*, a fat tome listing all the historical sources and books about the island, and a visit to the archives, which stocks a copy of all current books on the island, is impressive. Though the locally lionised writers are little known outside Mauritius, for the islanders they are a proud expression of Mauritian intellect. Robert Edward Hart, who died in 1954, was Mauritius' most renowned poet. Malcolm de Chazal, an extrovert flamboyant figure, is probably Mauritius' best known painter. His stylised flower prints in searing tropical colours are widely popular, and his work decorates new hotel rooms and now may be seen in batik work on silk scarves. Auguste Toussaint is an internationally known historian and writer renowned for his studies on the history of the Indian Ocean, who has been instrumental in setting up a Historical Association in the Indian Ocean countries. Marcelle Lagesse is the leading woman writer, whose novel *La Diligence s'éloigne à L'Aube* was published in Paris, and won prizes there. There are many more writers and poets, like K. Hazareesingh,

Regis Fanchette, Jean Nairain Roy, Pierre Renaud, Jean-Baptiste Mootoosamy, Loys Masson (who became famous in France) and the late Marcel Cabon. The remarkable thing is that so many people one meets are writing a book or paper about their favourite aspect of Mauritius. These may never be read outside Mauritius but will be avidly discussed and dissected there. There are no publishing houses as we know them, but the printers in Port Louis will print the manuscripts if the authors can raise the money.

Mauritius has formed the background for few internationally known stories. Joseph Conrad wrote of his unhappy love affair with a merchant's daughter in 'A Smile of Fortune', which has already been mentioned, and Sir Walter Besant, a teacher in Mauritius from 1861 to 1867, gave his novels *My Little Girl* and *They Were Married* a Mauritian setting. A French journalist of Mauritian descent, Paul-Jean Toulet, became head of the Fantaisiste school, writing poetry and novels inspired by Mauritius, but it was the sentimental French novel *Paul et Virginie* by Bernadin de Saint Pierre, published in 1773, which made Mauritius famous in France and a pilgrimage for visitors, French and even British. The novel provides a setting of pure nature, unpolluted by man, a garden of Eden in which a boy and girl grow up and eventually fall deeply in love. Virginie visits Paris to be educated and on her return her ship is wrecked on a reef on the north coast in sight of land and the adoring Paul. Virginie could have been saved but is too modest to remove her cumbersome clothes; she drowns and her body is washed ashore. Paul dies of grief soon after and the lovers are buried in the same tomb.

The idyllic purity of the novel made popular reading in the nineteenth century, and visitors arriving by boat would go to see the supposed tombs of the couple in the Pamplemousses gardens (see p 106). Today the fictional couple—though many people think they actually existed—are commemorated in names of streets, a restaurant and a hotel. The image of a harmonious life in a lush beautiful tropical setting is one Mauritius likes to dream of still.

There are several literary associations. The Académie Mauricienne, founded in 1964, produces a biennial collection of writings of the three Mascarene Islands. The Hindi Writers' Association was started in 1961 to promote Hindi language and literature in Mauritius. The Société des Ecrivains Mauriciens was founded in 1938. In addition, the parent countries send literary aid from outside: the British Council and Alliance Française have offices in Mauritius and the Indians and Chinese send dance and song instructors. An Indian Culture Association and Centre has been set up, and in the university the Gandhi Institute is being created as a school of Indian learning which it is hoped will attract students from all over the world. There are also current plans for an Institute of Bilingual Studies at the university which would attract students from Africa.

In addition to the Mauritius Institute the Vacoas House of Debators, founded in 1936, holds fortnightly meetings for debates, playreadings and speeches. European theatre companies once came and performed in Mauritius, but the expense is now too high, and theatre performances are now put on by amateur companies such as the Mauritius Dramatic Club, which can trace its English language productions back to 1823. A Mauritius Amateur Dramatic Club was formed in 1898 and performed until World War I. The present club was formed in 1932.

TIME OFF

The habits of indecision and procrastination, a preference for discussion to action, and remoteness from the pressures and fears of disaster felt in many other parts of the world, give Mauritius the atmosphere of another century. Everything is a little formal, a little faded. At the beginning of the nineteenth century Port Louis had an impressive social season, with parties, musical performances, opera, ornate carriages and

Paris dresses that all gave the appearance of dazzling sophistication and wealth. Something of that aura still hangs over a society seemingly unaware that the world has changed.

Women are still very sheltered. There are few in public life, though one or two have become town councillors and prominent social workers. Most of them spend their time looking after their large families, choosing new dress styles, having their hair done, meeting friends for lunch in Curepipe, and entertaining at home in the evenings. Women as daughters also live sheltered lives, their mothers watching anxiously over their choice of young men, trying to make sure they will suit the family. The young Chinese seem to have the most liberty, and may be seen in groups in restaurants and at the Chinese-run discotheques in Curepipe at night. Indian girls are well protected at home but are occasionally now seen alone with a boy friend at the beach or sitting in a park. A 'Patmos of idleness' was how General Gordon described the island in the nineteenth century, and to many it appears unchanged.

In the past soldiers, French or British, stuck at the Mahebourg end of the island, away from the Port Louis social centre, easily became bored, and John Alexander Ewart has remarked on 'a hot climate and want of occupation being rather conducive to quarrelling'. This led to much duelling in those early days. The same conditions still apply, but duelling and quarrelling have been replaced by endless discussion and argument on esoteric subjects. Mauritians keep busy labouring some local historical or geological point. The many clubs cover all kinds of learned subjects, and form a solid part of the social scene. Mauritians with lively minds need never be at home.

The cinema holds sway over the working population, and one sees whole villages streaming along the roads to form large queues outside them in the evening. The Plaza in Rose Hill is one of the biggest cinemas, and it is also used for public meetings and concerts, which are held fairly regularly. Mobile cinemas once took films round to village social centres, but the arrival of

Mauritius TV has made a TV set in the local hall the villages' most important possession.

Throughout all social levels association football is the most beloved sport. There are big stadiums in Rose Hill, Quatre Bornes and Port Louis, and the George V in Curepipe holds 15,000 spectators. As well as doing the British football pools, the island has its own league tables and a Challenge Cup played for in May and June; Mauritius takes part in the Indian Ocean Football Cup. Most big matches are held on Sunday afternoons. Many of the teams have racial affiliations, such as the Muslim Scouts and the Hindu or Tamil Cadets. Other teams have names like Dodos or Blue Ducks.

Athletics, volleyball, basketball and lawn tennis are other well liked sports. Though the projected Port Louis swimming baths have not yet been built, facilities for hockey, badminton and bowls exist in the Robert Edward Hart Garden.

From May to October the Champs de Mars in Port Louis is given over to racing. This is the highlight of the year, with all groups and strata of society mixing together in the small racecourse enclosure. Mauritius' big race is the Maiden Plate, run in late August. The racecourse, the oldest in the southern hemisphere, is crowded during the meetings, with much entertaining in the private boxes of the stand.

Some Mauritians are keen weekend mountaineers, walkers and climbers, but the main accent is on water sports. Now that more beach facilities have been provided with the coming of hotels, locals visit them on Sundays for lunch and swimming or fishing. The Morne Brabant Hotel has big game fishing facilities from October to March, a golf course and horse riding. Though there are a couple of small air strips, there is only one small plane on the island, probably because there is nowhere to fly to. The opening of the strip on Rodrigues, scheduled for 1973, may encourage more private flying. Sailing is popular, the season for the regattas held in Grand Port and Grand Baie running from July to October. Most of the weekend sailing clubs, like the Mauritius Yacht Club, are situated in Grand Baie in the north.

With the lure of wrecks, skin diving is actively pursued, organisations like the Mauritius Underwater Group and the Mauritius Underwater Fishing Association providing facilities. HMS *Mauritius* also has sailing and diving clubs and its own coastal club at Le Chaland in the south.

The wealthier Mauritians may ride, or hunt hare or wild pig the year round and deer from June to August. There is a golf course at the Naval and Military Gymkhana Club in Vacoas, a prominent meeting place for those with a British accent which also shows the latest English films. The Dodo Club is another old established social club, where members at dinner toast the club's set of dodo bones.

Most Mauritian evening entertainment is home-made. There are innumerable charity fund-raising and social welfare societies inexhaustibly arranging lunches, dinners and cocktail parties. Bazaars and exhibitions of handicrafts or paintings are well attended, and the more lowly hold frequent Fancy Fairs, the equivalent of 'bring and buy' sales.

In spite of their sociability Mauritians are early-to-bed persons. Dinner will be served around 7.30 at the latest. 'En famille', a large meal may be served in early evening when people are home from work and perhaps a hot drink will be prepared later on. The whole island (or 99 per cent of it) is in bed and asleep by 10.30, 11 pm at the latest. Punctuality is observed, though guests may—with genuine cause—plead losing their way the first time they visit a friend's house.

Until the development of tourism the Mauritian had little or no opportunity to live it up at night. Now more nightspots are being opened up and the younger generation is beginning to try them out. Even so, night life is very limited. There are one or two restaurants, mostly in the tourist hotels, with live or disc music to dance to, but no cabaret clubs. The pièce de resistance of Mauritian night life is the Casino in Curepipe.

Set in an old clubhouse, it has a smart modern interior—light fittings in the form of huge dice—and attractive Chinese girl croupiers. The main games are roulette, blackjack, chemin de fer, poker, and the Chinese quatre quatre (fan tan). Alongside the main gaming room is a restaurant where guests of the island's main hotels can eat on a voucher system.

Chiefly, however, pleasures revolve round family life. Weddings are gay affairs, church festivals of whatever faith are well attended, and births and deaths are equally the cause for get-togethers. To be seen to be playing one's social part is all important on this small stage, where everyone knows all about everyone else.

FOOD

In spite of the dominance of French culture, that nation's flair for cooking has impressed itself little upon Mauritius. In the hotels the visitor is carefully shielded from the local dishes by a stolidly international menu that uses imported vegetables and fruits rather than the delicious local products.

As we have said, there are few restaurants in which to try native cooking, and no prominent Indian restaurant at all. The only two roadside restaurants are La Bonne Chute in the Black River area, and the newer Le Grillon at Grand Baie in the north. Both are adjacent to petrol stations, but have charming unpretentious patio settings. Both advertise their weekend menus in the local papers.

The camaron or large river prawn is the most characteristic Mauritian dish, beloved by all. A weekend pastime among workers is catching it in the fast flowing rivers: a bamboo rod with a 4ft long noose on the end is held patiently in the water till the tail of a camaron is trapped in it, when the prawn is yanked out skilfully and put in a wire box to be taken home for lunch.

Camarons are usually grilled, and sometimes served with a

Page 89 (above) Bottle palms from Round Island growing outside the Sugar Industry Research Association buildings near Le Reduit; (below) although sugar is the main island crop, bananas are also grown

Page 90 The way of life: (*above*) the many rivers of Mauritius are used by the women to wash their clothes. In the background a typical pile of lava stones cleared from the sugar fields; (*below*) the Sega is the traditional creole dance of the island. Here village people perform it to traditional accompaniment of drums and metal triangle

cheese sauce. Most often they appear with a *palmiste* salad and *sauce rouge*, the other two notably characteristic dishes of the island. The *palmiste* is the soft centre of the heart of a coconut palm when it has grown 6–7 years. Removing it kills the tree, and it is, therefore, prized as a delicacy. It is most commonly served with a French dressing as a salad, or it can be pickled or cooked with a *sauce rouge*—which goes with virtually every kind of dish, even turning up as a tomato chutney to go with curries. It is made by heating thyme, garlic, onions and tomatoes in oil till the tomatoes burst and can be mixed to a smooth sauce. Salted fish is often served in this sauce.

With the poor quality of local meat, which is imported from Madagascar mainly, steaks are served with a hot pepper peri peri sauce. Inferior meat also means that Mauritius' main table delights come from the sea, though occasionally there is game and venison in the hunting season (June to August). Hare is widely served, as are small game birds. Sea food is magnificent, and one of the best places to taste it is the Touess-rok Hotel at Trou d'Eau Douce. Crabs, steamed or made into *bouillon de crabe*, are excellent. The locally termed *homard* is really a *langouste* (crayfish), often served in a soufflé with a *palmiste* salad. *Tec Tecs*, small clams, are served as hors d'oeuvres and so are *palourdes* (cockles) grilled with butter. *Tec Tecs* are also steamed in bouillon as a soup. Benetier clams are served with the meat baked and spiced in creole style.

In the eighteenth century Bernadin de Saint Pierre recorded in his list of sea foods '*L'huitre commune qui se colle aux rochers et d'une forme si baroque qu'on ne peut l'ouvrire qu'à coup de marteau*' (The common oyster that sticks to the rocks and which is of such an odd shape that it can only be opened by a blow of a hammer). The oysters are smaller and more humpbacked than those known in Europe, but they are cheap and taste excellent sprinkled with sharp local limes. Among the Indian Ocean fish caught and eaten is guele pave (silver bream) snapper, capitaine, sacre chien, and even cordonnier, which some say gives them nightmares. Octopus, known locally as *ourite*, is dried, then cooked over charcoal or even curried. Tuna is of course caught off the shores. Many of the sea fish are cooked in coconut milk.

Indian cookery has had its influence on the Mauritian cuisine. The national dish could be said to be curry, and dishes based on curried rice, such as pilaffs and briani, are common. The lobster is curried by the richer whites, and potatoes served on rice by the poorer. Characteristically served with the curries are little plates of chutney, not piquant but usually a purée of vegetables. Tomato chutney is most common but usually there is a little dish of cucumber as well. Tamarind, mint and arachide (or ground-nut) chutneys are also served.

Other common accompaniments to the main dish are *bredes*—various types of green leaves cooked plain or with a small quantity of fish or meat. These leaves can be watercress, Chinese cabbage, spinach or vegetable marrow shoots, or come from the mourounge or drumstick tree, solanum nigrum or deadly nightshade, amaranthus, chou chou or pumpkin. They are included in the daily diet of the majority of Mauritian country dwellers. Vegetables commonly used are the chou chou (pear-shaped baby marrows) and lalo or ladies' fingers (*okra*). Lentils and cassava fill out the diet and a dish of small sausages, fish or meat cooked with tomatoes known as *rougailles,* is often served with platters of black lentils. A cake-like mould is made from cauliflower. Chou chou is either served plain or with a white sauce.

In the absence of locally made cheeses, though some imported ones are available, dessert tends to be rather on the sweet and sticky side, such as the much loved Moka dessert. Gateaux and ice-cream concoctions are favourites, with sweets based on coconut. 'Napolitains' are shortcakes topped with purple or pink icing. Fruit flans are often served, apple tart French style is widely available, and there is sometimes a fresh fruit salad, particularly delicious in the summer months when the mangoes are ripe. It is strangely difficult, however, in restaurants to get the luscious fruits of the island, even though mangoes, paw paws and lychees abound in summer. Sweet, if small, Victoria pine-apples are sold, peeled and cut into spirals ready to eat from the street-seller's tray in Port Louis. In summer *longanes* are sometimes served; these are small round fruits covered with a hard woody shell, and are hard work since they have to be popped open with the fingers to reveal a pulpy white flesh somewhat like lychee in flavour but with hard bitter pips. Wild raspberries, Chinese guavas, custard apples, water melons, coconuts and bananas are also available.

Chamarel coffee is served in several restaurants, and tisanes of digestive herb teas favoured by the French will be made on request. Imported wine or chilled local Phoenix or Stella beer is drunk with the meal, the preferred hot weather wine being a chilled rosé. Surprisingly little rum is drunk by the more wealthy Mauritians, though it costs only 10 rupees a bottle.

They say the economy of the ordinary Mauritian is tied to the cost of *pommes d'amour*, the tiny cherry tomatoes that are grown and eaten all over the island. Certainly the Mauritian worker does not pay much for his lunch away from home. In the village and town streets, and in the Port Louis markets, are small stalls with glass cases containing cooked foods. Alongside is a small charcoal brazier on which a man endlessly, and al-most effortlessly, cooks the staple of life—the *gateaux piments*. These are whole green peppers in beignet-style batter, fried crisply. They are remarkably hot to the uninitiated though there is a cooler version in the *samousa*—little round crusty savoury cakes made from onions crushed with dahl. The worker buys 5 cents worth of butter, his *gateaux piment* costs another 5 cents, and then he adds a slice of bread and a sardine. The whole costs him about 15 cents, just a penny or two.

The office worker will probably bring his rice and curry lunch with him from home packed in a neat square *tente* with a fitted top. He will find a quiet place in the office and tradition-ally turn his face to the wall and eat quickly and quietly.

Others will buy fruit from the street stalls, and go to the market for thick *chapatti roti* with a 'chutney' filling.

Bread in Mauritius is excellent and readily available from corner stalls or markets—round *pains de maison,* square loaves bought almost by the yard, and French-style *batons. Dhall poori* is the name of more crumbly bread, made at a factory at the Trou aux Cerfs, Curepipe.

8   MOUNTAINS AND COUNTRYSIDE

MAURITIUS rises from coastal flats to high plateau. There eroded mountain peaks inscribe a gap-toothed circle round the island. The Piton de Milieu is its central point.

Baudelaire in his 'Invitation au Voyage' describes his impression of the interior:

> La, tout n'est qu'ordre et beauté
> Luxe, calme et volupté.

The overriding feeling certainly is of rich greenness, lush dampness and great fertility. Man, alas, is more fertile than the shallow soil! That is blanketed by sugar cane, a grey greenness softening and blurring the lines of the countryside, reducing roads to alleys between its high growth, and undulating in the sea winds. Cane fields cover over half the island.

Numerous rivers cut across it. These rivers, plus the artificial and crater lakes, which act as reservoirs, give the island good reserves of water, though the effects of drought are often felt. In the Plaine Champagne area the ground is sponge-like, releasing water throughout the year to feed the rivers of the south, which never run dry as they do in the north. The main watershed runs north across the central plateau for about 20 miles, and the rivers, therefore, flow to the east or west. Most are short and fast flowing, having gouged themselves deep

95

ravines over the centuries. Surprisingly for such a hot climate, there are numerous waterfalls.

The main note of harshness in this green island is struck in the giant piles of basalt rocks in the cane fields, particularly in the aptly named Plaine des Roches in the north-east. The rocks are broken-up pieces of the molten lava coating of the island. Some sugar estates build the rocks into rough walls, as they believe this keeps moisture in the soil.

When the island's railways were closed in 1964, the sugar estates, the principal users, agreed to help pay for a system of roads, and the island is well covered. They have tarmac surfaces, and suffer in the cyclone season; but new ones are still being created, and old ones, like the original Port Louis-Mahebourg road, are being reopened. Roads on sugar estates are privately owned, and permission must often be sought to cross them, though there is always an alternative public road. Private property is a well preserved and respected concept in Mauritius, and 'no trespass' signs are liberally sprinkled alongside the roads.

Lack of signposts, no road-numbering system and the difficulty of obtaining accurate up-to-date motoring maps make crossing the interior of the island more complicated than it need be. Nor are there any little restaurants or cafes to refresh the traveller. In 1972 a French company surveyed the island, and it is producing maps in association with the Michelin company. There is a scheme for signposting under the direction of the Development Works Corporation (see p 155), but this is still awaiting government approval. There is also the problem of which language to use—French or English—or to produce signposts in both.

THE MOUNTAINS

In Mauritius the flat plains about them, and the clarity of light, add drama and stature to the mountains. The eroded lava has been weather-worked into such shapes that names and legends are easily attached to them. In the old days, when the mountains and surrounding foothills were liberally covered in forests, the peaks were the hideouts of escaped slaves known as maroons, until Labourdonnais checked their numbers. Even now, however, their heights provide some form of escape for energetic Mauritians. Climbing the now uninhabited mountains became popular in the nineteenth century after John Augustus Lloyd, an engineer, climbed the reputedly inaccessible Pieter Both in 1832 with three companions, though there are records stating it was climbed by Claude Penthe in 1790. The crumbly state of rock makes some mountains dangerous, as it is impossible to put in pitons, and such conditions makes Le Morne and Rempart unsafe to climb. Sound advice from a man who has climbed them all is contained in a small book, *Climbing and Mountain Walking in Mauritius*, by Alexander Ward; he is head of the Special Mobile Force, members of which are trained and toughened in regular mountain climbs.

Pieter Both (2,699ft) is perhaps the most impressive of Mauritius' mountains, with a huge boulder poised on its tip, seemingly about to tumble to the valley below. The French say the day the British leave the island the rock will fall. As it remained in place on Independence Day, 12 March 1968, local legends say British influence is still strong on the island. This 30ft rock has made climbing Pieter Both intriguing and difficult, and the mountain has claimed several lives. Lloyd's expedition, however, did it in comfort, taking up smoked salmon and good wines. The climbers managed to erect a flag on top of the rock, and then drank a toast to King William IV. The night was so starry, and the lights dying out in the towns below looked so entrancing, that the party stayed the night on top of the mountain, 'having lashed Phillpots, who is a determined sleep-walker, to Keppel's leg'. The indefatigable Nicholas Pike climbed this and other mountains. He describes the parties of British navy and army officers who climbed it in 1848, 1858 and 1864; the last party left a visitors' book in a tin at the summit, and Pike remarks: 'I do not think the book requires to be a very bulky one.'

The mountain, now tamed, is not difficult, but it is recommended that one skilled climber with ropes accompanies each party. The mountain takes its name from an early Dutch governor, but a British governor, Sir John Pope Hennessy irreverently called it 'The Queen [Victoria] in her coronation robes.'

Pieter Both lies at the eastern end of the Moka range, the hills which curve behind Port Louis. To the west of it is Virgin's Peak and to the south-west Le Pouce (the Thumb), 2,661ft high, whose peak is shaped like a raised thumb. This is relatively easy to climb, mostly by an old drove road leading over the shoulder of the mountain from the south to Port Louis. After passing through part of the original forest, one has a steep scramble to the top, but the rewards are a superb view of Port Louis and the island. Alternatively, Priest's Peak and 'the window', a square cut-out looking down over Port Louis, are easy to get to and can be climbed in under 2 hours. Pike, of course, climbed Le Pouce:

> The path to the summit is narrow and steep, a mere scramble up rocks; and when there we found only a little plateau about ten feet square. The whole island lay around us; and it was a glorious sight to look down on it from that giddy pinnacle, so calm and lovely in the far distance, and not a sound saving our own voices to break the silence. . . . The city lay at our feet in a northerly direction; the plains of Pamplemousses, and Rivière du Rempart, to the NE, were green with waving canes; and the large plantations many of them over 1800 acres, looked only as so many cultivated gardens. The Moka and Black River districts to the W presented a similar scene. . . . The Latanier and other rivers in their serpentine course meandering slowly to the sea appeared as silver lines intersecting the country. The tracks of the railways were just visible, and as a train passed, no sound reached us; but as the iron horses rushed puffing along, they seemed like children's toys rather than monster engines . . .
> It was a new view of the city to me with all its surroundings: the

harbour and its forest of masts, the wreaths of foam marking the coral reefs; the forts; and the broad expanse of the Indian Ocean, all glittering in the brilliant tropical sunshine.

For the less energetic, Signal mountain, with its now abandoned signal station on top, also provides a good view of Port Louis. It lies at the end of the Moka range as it curves around practically to the sea.

Southwards another group of mountains can be seen particularly well from the Curepipe/Rose Hill area. Sunset on the Corps du Garde mountain (2,558ft) gave the name to the town of Rose Hill. Due south of Corps de Garde are the Montagne du Rempart, which from the coast road resembles the Matterhorn in silhouette, and the Trois Mamelles, which look like an upturned cow's udder. It is a hard climb of 2,187ft to the central peak of the latter.

Between this mountain and the high Chamarel hills of the south-west corner of the island is Mauritius' highest mountain— Black River Peak at 2,711ft. Expert climbers say it is, however, more of a long walk than a testing climb. In this area wild pig, deer and monkeys abound, and the views show forest ridges, waterfalls, and, to the west, salt pans near Petite Rivière Noire Bay. There are also good views of Le Morne Brabant mountain (1,824ft), whose massive square-headed form falls sheer away to the sea, forming a small peninsula. In sight far down the coast as one proceeds south from Port Louis, Le Morne is more impressive after one cuts across the Chamarel peninsula and starts heading east along the south coast. It gives one the impression that a piece of Ireland has strayed into the tropics. Le Morne has a deep ravine crossed by a footbridge that was erected in 1835, when a policeman was sent to tell escaped slaves hiding there that slavery had been abolished. The slaves, either disbelieving him or having too many crimes on their consciences, threw themselves to death off the mountain. The Cambier family, which now owns a 2,500 acre estate on the mountain, found their skulls decades later.

Le Morne is being developed as a tourist centre by the Cambiers. Bungalows are being built into the hillside, and hotels and recreational amenities will be added to those already in existence. A safari club will be set up on the mountain, with its soaring views of the two coasts, and it is hoped to introduce some animals from Africa and to set up stag-hunting areas.

The other mountains of Mauritius are less known. In the east the Grand Port range divides the river valleys around Mahebourg. Just behind Grand Port itself is Lion mountain in the Bambou range, named for its crouching lion outline, just as Le Chat, further north-east in the range, appears from the coast road to have the figure of a cat watching a mouse on its summit.

North and parallel to the Bambou range are the Blanche and then the Fayence mountain ranges. In approximately the centre of the island is the Piton de Milieu, a low cone attractively set by a reservoir of the same name; and almost due north are the La Nicolière mountains guarding La Nicolière reservoir and having Mount Nouvelle Découverte as their highest point.

THE FORESTS

Although the huge slow growing woods have now gone, there are still extensive patches of the original forest preserved, as in the Machabee area, which afford pleasant regions for walking and natural history study, though as yet they have been little publicised. They are under the control of a Forestry Department, the oldest in the Commonwealth and originally set up by the French in 1777, headed by a Forestry Conservator. Experimental gardens and offices are situated next door to the Botanical Gardens in Curepipe. In addition to clearing areas of forest for replanting as tea estates, or with conifers, the Department is encouraging tourists to see more of them. Picnic areas have been formed and clearings made at beauty spots, and visitors can drive through the forests on narrow tracks. Several such tracks and viewing points were created for the 1972 visit of Queen Elizabeth II, including a sight of the superb waterfall area in the Black River gorges.

In this part of the island, it is possible to drive along the parallel spurs of ridges above the gorges to various look-out posts. A most charming spot is the look-out at Mesliers Point over Bel Ombre in the south. Here the old funeral coach from the defunct railway system has been set up and can be used as a picnic shelter. Inside, a table has replaced the former area for the coffin, but the solemn leather-covered chairs for the chief mourners still remain, as do the leather benches for mourners of less importance.

Information on visiting the forests can be obtained from the Forestry Department, and the keys for paths and picnic places are obtainable from foresters on the spot.

The main area of indigenous forest is in the south-west corner of the island. From Curepipe it is a few minutes' drive to the Plaine Sophie, which is planted with pines. On the road is a cairn erected by the Ancient Monuments Committee to Matthew Flinders, the explorer, commemorating the spot where he made an enforced 6 year stay as a guest at the farm of Madam D'Arifat. On the road past the pine-edged Mare aux Vacoas (see p 103) is a Forest Sub Station with beautifully kept gardens, where information on the forest paths can be obtained. There are walking trails and *brisés* (open paths), maps of which can be obtained from the Forestry headquarters in Curepipe.

On the other side of the Mare Longue reservoir lies the dense Machabee forest, backed by the Piton de la Rivière Noire. In this forest one can see fine examples of the natte tree. Leaving the surfaced road, one follows the forest tracks over the Rivière des Aigrettes, which feeds the reservoir. One can either enter the forest area of the Black River gorges to the west or first cross the heathlands of Le Petrin, past clumps of tropical heathers. Much of this area is being cleared for replanting.

The Black River gorges area is one of the most beautiful interior parts of the island. Tracks run along the tops of thickly forested hills that divide the rivers, and 300-year-old ebony trees may be seen. Viewpoints have been cut so one can see down into the plunging gorges to the falls and rivers below. In the skies above tropic birds wheel displaying their wide white-feathered tails.

At the westerly tip of one of the tracks, 2,000ft above sea level, is a kiosk pavilion for picnics and an exhilarating view over the cliff-like hills towards the Brise Fer mountain and Tamarin Bay. The Black River peak is on the left, and it is also possible to see the Rempart and Corps de Garde mountains to the north. Tracks also lead to the lovely Tamarin falls and reservoir. From here, without the clustering of houses and humans that characterises much of the island, one can stand warmed by the sun glancing down past the black silhouetted mountains to the mirror flat sea crinkling round the reefs. From such vantage points Mauritius is timeless.

Another westerly branch off the main road at Le Mares, where there is a forest office in an old railway carriage, leads across the more open grassy Plaine Champagne. From here little side tracks lead to viewpoints whence one may see the Black River gorges and Black River peak over a 400ft sheer drop to a cascade. The road west has now been continued so that it is possible to drive down curling bends through the trees to Chamarel and the coast near Le Morne.

LAKES AND RIVERS

Les Mares has a branch road leading east to Grand Bassin, from which one may travel either to the Kanaka crater or straight on to the tea estate area of Grand Bois, whence main roads lead back to Curepipe through Nouvelle France, a village notable for the roses, hydrangeas and artichokes growing in the gardens. Grand Bassin is one of Mauritius' crater lakes, with a peak on its south-west side. Nicholas Pike after seeing it was disappointed: 'I think it has been greatly overrated.' It has become renowned for the annual Maha Shivaratee Hindu celebration. Like the other lakes in this area, Grand Bassin makes a good picnic spot.

The Mare Aux Vacoas is Mauritius' largest reservoir and a road encircles it. It is not always possible to get close to reservoirs and the Piton de Milieu reservoir and Eau Bleue, also in the centre of the island, can be viewed only from a little way off. La Ferme reservoir lies between Rose Hill and the coast. The east side of this natural reed-shored lake noses into a V of high hills forming the Corps de Garde and Mount Saint Pierre. One can drive along the east shore of the lake to the hydro-electric power station; the low scrub trees make a good picnic spot, with sun reflected off the steep grey stone hills behind, over which the huge water pipes run.

The other notable stretch of inland water is La Nicolière in the Villebague area. This is an artificial dam with a road over it, from which one can see the stone channels through which the rainwater courses from Curepipe, 10 miles away.

There are about twelve major rivers, the longest being the 21 mile Grand River South East, followed by, with lack of imaginative naming, Grand River North West. In the months after Christmas the rivers are specially full and waterfalls are at their most impressive. The rivers are daily full of women, their saris hitched round their knees as they beat and scrub the family washing, spreading it to dry on pandanus plant spikes. Mark Twain remarked caustically of their activity: 'They should find another way to break the stones, the clothes will wear out first.' At weekends the rivers are lively with boys fishing or hunting the camaron prawn.

The splendour of the rivers and falls slicing their way through the earth's crust is dramatically demonstrated from the gardens of Le Reduit or the Botanical Gardens in Rose Hill, which give a fine view of the Grand River cascade. More easily accessible than the Tamarin Falls are the Rochester and Chamarel Falls. Signs from the village of Surinam near Souillac on the south coast lead one to Rochester Falls, the last part of the way across a grassy track between sugar cane. One can cross the top or slither down a steep bank to have the best views of the falls, which have worn the rocks into vertical columns.

In the south-west corner the Chamarel Falls and curious Coloured Earths may be reached by following a steep road up from Case Noyale, which is on the coastal road. The road to Chamarel Falls gives one superb views down over the coast and Le Morne mountain, glimpses of stags in the game reserves below, and monkeys gibbering along tree branches. The Falls and Earths are on private estates and a charge of 1 rupee per person is made to enter. The track is rough through the cane fields but the view of the Rivière du Cap Falls, a 300ft plunge, is excellent. A mile or so further on one finds a little hillock of coloured earths, a weird mound of soil in vivid stripes of purple, red, blue, yellow and rust, isolated amid cane and forest. There is no other soil like it in the island and no one is quite sure of its geological explanation, though the earths are said to be volcanic cinders eroded bare and coloured by oxidation. The bands of colour undulate across the lava dunes and are best seen with the sun on them. The colours have not mixed together as they are of different densities.

The area can also be approached from the south coast at Choisy, the road climbing over the Chamarel mountains, named after a French army officer. The area gives it name to the locally grown coffee, which can sometimes be obtained in the Port Louis restaurants. It is slightly sweet, as the beans are roasted with sugar. Less acknowledged products of these hills are *gandia*, the local name for marijuana, whose growers are rooted out in raids by helicopter, and an illegal rum called *tilanbic*.

HUNTING

On the grass verges of the road which rises to Chamarel from Case Noyale are several tall chairs like those of tennis umpires, built of wood with a little platform at the top. These are *miradors*, used by the huntsmen during the season of deer hunting from June to August known as 'La Chasse', and an important part of Mauritian social life. Pike noted:

> Deer hunting in Mauritius is quite an institution and is popular with both Europeans and Mauritians; indeed with the latter it amounts to a *grande passion*. When a *chasse* is proposed, no need then to complain of the ordinary indifference or laziness; on the contrary, every one is roused to no end of activity. The hunting season begins on the 15th of May and terminates at the end of August . . . During the season there is a *chasse* about once a fortnight, and I have seen as many as thirty six deer killed in a day.

A plan is being mooted between herd owners and the sugar estates to rear the deer on a commercial scale, and experts from the FAO are studying the deer on Le Morne estate. In 1972 it was estimated there were 20,000 in Mauritius, and this number could be raised to 60,000 in 4 years. Such a plan would involve changing the law to permit killing deer all the year round.

Traditionally La Chasse starts in June and continues till August. (Wild pig and hare are hunted year round.) The occasion is more social than sporting. The deer are beaten past the guns of the men in the *mirador*. After the hunt there is a massive feast given by the owners of the land for their friends. Even at Le Reduit, with its forest ravines, former governors would give hunting parties for their friends. In the *Almanac of Mauritius* of 1854 Mr Draper-Bolton describes a typical hunting party:

> Each guest brings his own servant and quota of provisions, liquid and solid, and also some sort of bedding; and those who

have good dogs are expected to bring them. Each chasseur is provided with his gun, generally a double-barrelled smooth bore. After an early breakfast, they sally forth, the post of each person being agreed on, and when all are at their stands, the first relays of dogs consisting of a few couple is thrown off, the others being held in leash ready to slip when the deer passes in the direction in which the respective relays are posted. As the deer passes the stands he is fired on successively till a well-aimed shot brings him down. Two or three are frequently killed in a day; and the number is now so great on well preserved chases that they will soon require thinning . . . Although the deer are so plentiful it is easy to conceive that in this mode of hunting many of the hunters never see the game. Many are passionately fond of the sport; but many others get disgusted at standing alone, often in wet and chilly weather, hour after hour.

Apart from visiting the museum in Port Louis (see p 69), where a comprehensive collection of stuffed fauna of the island is housed, seeing the wild life of the island is a hit and miss affair. There is one bird sanctuary, owned and created by a Mr Lenferna on the coast road near Yemen. The cages of birds are well worth the stiff entrance fee, though it is closed on Sundays and public holidays.

PAMPLEMOUSSES BOTANICAL GARDEN

The pride of the north is undoubtedly Pamplemousses Botanical Garden. Nicholas Pike wrote:

At the distance of seven miles from Port Louis, in the district of Pamplemousses, are the celebrated Botanical Gardens, founded by M. Poivre in 1768. These gardens have been from time to time replenished from the various botanical gardens of Europe, Cape Town, Australia and India and now form the special attraction of the colony.

The gardens, which once were rated the third best in the world, were originally created by Labourdonnais, who wanted to experiment with new crops for the island, and so started plantations round his home of Mon Plaisir. It was neglected by his

Page 107 (above) Indian procession for the Maha Shivaratee festival, carrying decorated arches known as 'kanvar' and passing through the centre of Port Louis on their way to the Grand Bassin to fetch holy water; (below) the Central Market in Port Louis, the largest in the island. On the pavement are the small, sweet Victoria pineapples grown on the island

Page 108 Communications: (*above*) Mauritius' motorway. The dual carriage road runs from Port Louis to Vacoas through beautiful scenery; (*below*) Plaisance airport in its early days; now the rapidly expanding airport is served by most major international airlines

successor, Governor David, who started plantations at Le Reduit, with Aublet as his expert. When the Intendant Pierre Poivre, who had brought prized pepper plants from the Dutch Indies to plant in Mauritius, quarrelled with Aublet, who was accused of killing the young pepper plants by pouring boiling water on them, Poivre started his own work at Pamplemousses around 1755. Philibert Commerson, who went round the world with Bougainville, was a naturalist who worked with Poivre, and a monument to him is erected at Flacq, where he died in 1773. The Pamplemousses gardens flourished and developed nursery plants for growers in the island, 50,000 being distributed in 1865.

Today Pamplemousses has lost much of its glory, but it still has much to impress those interested in tropical plant life, and is a cool haven to visit on a hot day. The heavy ornate gates and railings, which won the first prize for wrought-iron work in the Great Exhibition of 1851 in London, were the gift of a M. Lienard, commemorated in an obelisk in the main avenue. The inscription also states that a gift of a plant is worth more than a gold mine, and lists the people who did much to create Pamplemousses, including Poivre, Cossigny, Commerson and Aublet. At the end of the avenue is a little palm arcade shading stone benches, by legend the place where Labourdonnais went to write his reports.

In the nineteenth century Pamplemousses was a place of pilgrimage for devotees of *Paul et Virginie*, for it was thought that the sad lovers were buried there. Their 'tomb' is still to be seen, set at the end of a green hedged alley, but now obvious as the base pedestal of a statue to Flora. Now the star attraction is the long lily pond, in which float Victoria amazonica lily pads large enough to act as a boat for a small child.

The nineteenth-century house, often thought to be the French governor's house but in reality the botanist's house and herbarium, stands at one end of the gardens. Nearby is a reproduction of an old sugar mill 'factory', showing how wooden cog wheels were used in the crusher, how the sugar was heated in pans and how the crystals were packed into loaf-shaped wooden moulds. Also nearby is an enclosure of giant tortoises.

Gourami fish swim in a large canal encircling a palm island, and everywhere there are examples of different kinds of palms, including the bottle palm from Round Island and the dictoyosperma alba, which has a net-like tracery around its seeds. There are talipot palms from Ceylon, striped bamboos and some latanya 200–300 years old. Junipeis, araucarias and mahoganies abound. Seedlings of eucalypti and filaos are sold at low rates to the public. Across a ravine over the 'Bridge of Sighs' lie the foundations of the former English botanist's house, in which the 'bathroom' comprised a platform beside a stream and a slave with buckets of water. This part of the garden contains the bottle palms from Round Island, and a century palm which flowers once in every 100 years (this one last flowered in the 1930s).

Just outside the garden gates is the cemetery of the village of Pamplemousses, and an ornate collection of tombs of some of Mauritius' leading families. Among the elaborate wrought-iron crosses and high sculpted urns is the grave of Monsignor Antoine Buonavita, Napoleon's almoner on St Helena, and the huge square tomb of Adrien D'Epinay (d 1840), a reforming politician who founded *Le Cerneen*.

SUGAR ESTATES

In the awful days of the malaria plague of 1867 many crawled into the sugar cane to die. British novelist Sir Walter Besant, then a teacher at the Royal College in Curepipe, wrote: 'When the canes were cut, dead bodies were found of poor wretches who had crept in to die at peace under these waving plumes of gray.'

From the roads the canes may hide the workers; they also

hide some of the most magnificent old houses of the island. Permission to visit sugar estates may be obtained from the estate managers. Among the most beautiful estate houses is that of Villebague, one of the oldest estates, founded in Labourdonnais' time. Superbly laid out gardens characterise many of the estates: for instance 'Solitude', lying on the main road from Port Louis to Trou aux Biches and by a lake, gives a cool green lawn welcome; and Riche-en-Eau, in the south-west, has a splendid avenue of palms leading to the terraced house, with open vistas of the rivers flowing through the property. Aptly named, Riche-en-Eau has miles of rivers and streams, and the estate manager has created a delightful tropical garden in one of the river ravines, with a splendid array of tree orchids, a small aviary and a camaron farm, with water pens in which the river prawns are bred and kept. This estate has its own shop and recreational area for its workers—typical among Mauritius estates—and has one of the largest factories. The largest sugar factory of all is 'Fuel' (Flacq United Estates Ltd) near Flacq.

9 ROUND THE COAST

THE 100 mile coastline of Mauritius, well endowed with gently shelving white coral sand beaches, could be her economic salvation in tourism earnings. Most of the coast is sheltered by reefs, with wide lagoons for water sports and fishing.

The long sand beaches are mostly backed by feathery, dark green casuarina pines, known locally by the Portuguese name of 'filaos'. On the north-east coast the beaches are markedly empty and tempting for beachcombers. Charles Baudelaire was an early visitor to succumb to their charm when he wrote in 'Parfum Exotique':

> I see happy shores unfolding
> Dazzled by the fires of a monotonous sun;
> A lazy island where nature gives
> Peculiar trees and tasty fruits
> Men whose bodies are slim and strong
> And women whose eye is astonishing in its frankness . . .

Among the slimmest and strongest are the fishermen in their pirogues, gliding through the lagoons or working with hand-nets. In spite of illegal dynamiting of fish within the lagoon, which has destroyed much marine life, Mauritius is rich in fish. Japanese tuna fleets are, by agreement, based in Mauritius, and they take their catch to Japan for canning.

The sea round the island is influenced by the currents of the South Equatorial Stream in the surface waters bringing warmth and flying fish from the Central Indian Ocean. Deeper down, the Aghuelas Counter Current flows up from Antarctica, bearing with it rich plankton and sea salts which occasionally well up like rivers round Mauritius and the Cagados banks to the north. This provides nutrition for the big fish like tunny, marlin, wahoo and dorado, which normally stay deep down but near Mauritius can even be seen by the skin diver around the coast.

Big catches of blue shark, marlin and tunny have been recorded, and international competitions are held every year. In 1967 a world record catch of an 1,100lb marlin was made off Mauritius. Within the reef, flounders, tropical mullet, red rock cod and crayfish, octopus and black carp can be seen while one is snorkelling. Mauritians love spending weekends and evenings fishing in the shallows of the reef. In 15–20ft of water jack fish, parrot fish, rock cod, sea bass, and crayfish are found, the Black River and Morne Brabant coast areas being especially good hunting grounds.

TREASURE HUNTING

The coral reefs that so delight the visitor on a glass-bottomed boat trip have caused a spectacular number of wrecks around the coast over the centuries. Blown off course by cyclones or ignorant of the channels, ships have had little chance of escape. In 1773 the *Ambulante* was blown out of Port Louis harbour and down the coast in a cyclone, eventually grounding in the middle of the Morne reefs, where the pass is now called Ambulante; and in the same year thirty-two ships were destroyed in Port Louis harbour in another cyclone. In the eighteenth century French corsairs sank British East Indiamen laden with goods going to and from India.

It is not surprising, therefore, that treasure hunting is a local hobby, and with some a full-time occupation. Perhaps the best known treasure hunter is Esperance Becherel, who lives with his Swedish painter wife in Baie du Tombeau, just north of Port Louis. For 6 years he has been excavating a flooded site nearby, where he believes £10 million of corsairs' treasure from the seventeenth century is buried. He has found several tortoise-shaped stones which he thinks are markers, and a wall that may lead to underground vaults in which he believes each ship stowed its bulkier treasures, flooding them for protection. Even if this treasure proves elusive, Becherel has several other sites in mind.

Just a few cannon and some coins have been recovered from the ocean, but the positions of many wrecks off the coast have been plotted. The Mauritius Underwater Group, formed in 1964, is a thriving scuba club that has explored many of the wrecks. Its members often work in co-operation with the HMS *Mauritius* diving club.

One of the located wrecks is the famed *St Geran*, which foundered in August 1744, when carrying machinery for the island's first sugar factories, on the reefs off Amber Island (Isle d'Ambre) in the north-east with heavy loss of life. She gave Bernadin de Saint Pierre the idea for his heroine's fate in *Paul et Virginie*. The *St Geran*'s bell has been brought up and is now in the Mahebourg Museum, and some silver pieces of eight have also been recovered.

The *Speaker*—the pirate treasure ship of John Bowen and Nathaniel North—sank in January 1702 and is thought to lie near the Iles des Roches on the east coast. Surcouf, the most famous local desperado, is said to have dumped the rich treasure he had captured from HMS *Kent* in Port Louis' harbour when ordered to hand it over to the French government in November 1800. Port Louis' harbour is littered with wrecks, mostly caused by cyclones. Among those still visible are the *Taher*, wrecked off Fort William in March 1901. Partially salvaged, she lies in 25ft of water not far from the *Tayeb*, wrecked on the reef in the cyclone of February 1972. The naval battle of Grand Port in August 1810 led to the sinking of two British frigates, *Magicienne* and *Sirius*. Both these wrecks are known and the Mauritius Underwater Group will escort divers to them. Many

objects recovered from the ships are on show in the Mahebourg
Naval Museum.

SHELL COLLECTING

Another, more common, way of treasure hunting in Mauritius
is to collect shells. The island is rich in varieties, harbouring
some of the most rare and valuable shells in the world, such
as several varieties of conus (with a world record of a conus
rattas, a conus elytospira and a conus geographicus of over
7½in), lambis violacea, and cyproea onyx-nymphal. The
thickly ribbed imperial harp shell is indigenous to Mauritius,
and the single harp is fairly common still. A good specimen of
the double harp will fetch 500–600 rupees.

Shell hunting in Mauritius did not become popular till about
10 years ago, and interest was largely aroused by Frederic
Descroizilles and his wife, who took their family out collecting
every weekend. Now the family has a priceless collection which
is on view to visitors at their home in Curepipe, where they
also sell shells and shell souvenirs. Madame Descroizilles makes
jewellery from shells, slicing them through to produce pale
pink earrings. Plans are being made to move the collection to
a more spacious museum setting.

Bernadin de Saint Pierre was the first to list Mauritian shells,
in 1773, when he recorded forty-eight different species in his
book *Voyage à l'Ile de France*. Another early shell hunter was
Elize Lienard, who catalogued 880 types of shell, and whose
collection forms the nucleus of the Port Louis Museum display,
at which all the world's major families of shells are represented.

Shell collecting in Mauritius has not remained a charming
amateur beachcombing hobby, alas! It has been exploited
commercially, threatening the extinction of certain species.
The fishermen, lured by the reward of quick rupees from tourists, bring back shells with them from their fishing trips, setting
up stalls along the roadsides near the hotels or in their own

homes. They also sell to the professional dealers, who export to collectors all over the world and have their shell shops in Curepipe and the hotels. The village of La Gaulette in the south-west has several shell shops run by local fishermen. Their wives run them during the day and sell the cleaned and polished shells, or blowfish made into lamps and small stuffed turtles.

Frederic Descroizilles is worried about the disappearance of species of shells within a few years, declaring that unless the government legislates 'there are only two years left for the shells'. *Le Cerneen* of 19 February 1972 warned of the need to tighten up restrictive export laws, stating that the unreasonable exploitation was rapidly bringing about the disappearance of shells. A law already prohibits the export of helmet shells used for cameo making to Italy. These shells were once commonly used in Mauritius as doorstops. Cowrie shells, another common kind, were used in every home as a darning 'mushroom', and even now on beaches like Belle Mare small ones can be found in a myriad of soft colours. Conches and cones are other common types. Of the lambis type of conches, the giant spider conch and the arthritic spider conch are common, and among the most delicate and beautiful of shells are the paper argonaut and comb of venus. Though the casual visitor will need to search hard for onshore finds, the male cowrie may be discovered in the Chaland area; cones at Trou d'Eau Douce, Black River and Mon Choisy; augers and mitres on the northern edge of the Morne reef; and olives at Flic en Flac, Chaland and Grand Gaube.

The cone shells are delightful—the long thin lettered cone, its creamy shell dotted in black, or the textile cone, with its delicate tracery patterning. There are six deadly cones to beware of; they should be picked up from the crown or thick end, put quickly into a container and stored safely for 2 days until they are dead. The injected poison from the sharp end is fatal within $2\frac{1}{2}$ hours, and no cure is known. The conus geographicus now on show in London's British Museum killed its Mauritian finder. The deadly types are conus geographicus, marmoreus, textile, tulipa, aulieus and rattus. Starfish are also common, including the crown of thorns (star achantaster), which is destructive to coral reefs.

OFFSHORE ISLANDS

Apart from her inhabited island dependencies (see Chapter 14), the coast of Mauritius has many small islands, though some are little more than sandbanks. Certain of them are privately rented but others, like those on the east and south coasts, act as weekend escape spots from the crowded community life of Mauritius.

The largest islands are in the north-west, visible from the coast around the Grand Baie area. The nearest is Gunner's Quoin, its westerly leaning silhouette with steep 500ft drop resembling the rest of a gun. North-east of Gunner's Quoin is Flat Island, which has a lighthouse and former quarantine station (mainly for Indian immigrants), and its own small boat service from the Port Louis docks. Before the railways were opened in 1902, mules were bred there for transporting the sugar from field to ship. Anyone visiting this island should report to the Harbour authority before doing so. At low tide Flat Island is joined by a reef to Gabriel Islet on its east coast. Another Flat Island neighbour is Pigeon Rock, whose dramatic appearance is described by Nicholas Pike:

> Pigeon Point, whose top is white with guano. The sides appear almost perpendicular but could, nevertheless, be easily ascended if a safe landing could be secured. When we saw it, the waves were madly breaking against it, throwing up columns of spray, and the current swirling rapidly round its base. This is an isolated basaltic cliff about half a mile from the shore, and rises to the height of 110 feet; the top appearing nearly level. On the shore opposite the Columba (Pigeon) a ridge of detached basaltic rocks extends, piled up irregularly, but all resting on coral.

117

MAURITIUS

The most interesting of the islands are Round Island and Serpent Island. Both are misnamed: the two species of unique indigenous snake live on Round Island, which is *not* round in shape; and Serpent Island *is* round but has no snakes, being a bird sanctuary. Round Island is about 14 miles north off the Mauritius coast and the cone of Serpent Island is further north still. Though boat trips are run to the islands, the sea current often makes it impossible to land there. November is the best month to try. The tufa has been eroded away underneath by the sea, forming a kind of Christmas tree shape. Locals say if you can see the 'trunk' of the island, then it is safe to go in and land. Kidney-shaped Round Island has steep hillsides with bottle palms. Goats and rabbits have been introduced to the island, and with natural erosion and destruction there are moves to try and raise funds to recreate forests there and protect the remaining wild life.

The north-east corner of Mauritius is deeply indented, bays enhancing dozens of small islets. Lying off the coast of the Poudre d'Or area but inside the reef is the Ile d'Ambre, so called after the ambergris which was found there when blue whales inhabited the seas around Mauritius.

A favourite of young Mauritians who want to be alone is the Ile aux Cerfs in the estuary of the Rivière Seche. There are several separate islands in this group—Ile de l'Est, Ile aux Lubines and Ile Forency. On one of the tiniest rock islands is the Touessrok Hotel and restaurant, and after lunch it is pleasant to hire a local pirogue and paddle across the shallow lagoon to the other islands. They are deer preserves thick with filao trees and grass in which hares abound, and edged with shell-strewn beaches studded with rocks. It is said that the Dutch once kept deer on the islands for food, but now the deer keep shyly in the shadows. The days of peace seem numbered, however, with plans in being for a hotel on Ile aux Cerfs.

Southwards the islands become smaller and flatter, little more than raised portions of the reef, which swings out about

4 miles from shore as it curves across the entrance of Grand Port Bay. Just south of the Iles aux Cerfs are Ilot de Roches, Ile Flamand and Ile aux Oiseaux out on the reef edge. Opposite Grand Port Bay are grouped together Ile Marianne, Rocher des Oiseaux and Ile aux Fours. Ile Marianne, a favourite with shell hunters, is named after the seabirds that inhabit it.

The other group of islands, with their treacherous shoals and misleading channels, at the entrance of Grand Port harbour were the undoing of the British frigates in the battle of Grand Port in 1810, when those ships foundered on the reefs. The Ile aux Fouquets has an unused lighthouse. The Ile de la Passe, captured by the British before the 1810 battle, marks the entrance channel with the Ile aux Vaquois. After the British occupation of Mauritius in 1810 fortifications were built on the Ile de la Passe, and the names of the British soldiers garrisoned there—and no doubt bored to death—can still be seen inscribed on the walls.

Within the port area are other islands, with more grass and trees, and often a weekend *campement*, like Mouchoir Rouge off the Pointe des Regattes and Iles Singe and Chat off Vieux Grand Port. The largest, and a favourite weekend spot, is Ile aux Aigrettes, south-east of Mahebourg. Des Deux Cocos shelters in the lee of Blue Bay, to which people may sail from the Le Chaland Hotel area.

The rest of the south coast is bare of islands until one reaches Le Morne peninsula. To its south lies Ilot Fourneau, and north of it is the large and privately rented Ile Morne. Permission must be obtained from the guardian before one may visit the latter. The island is long and about $\frac{1}{2}$ mile wide, curved and planted thickly with filao and coconut palms. The Ile Morne is set in one of the most extensive and beautiful of the lagoons, on average only 6ft deep, with views east to the coast of the Black River gorges and the Black River peak, north to the misted outlines of the Tamarin mountains, and south to the brooding overhang of Le Morne mountain.

MAURITIUS

The fishermen of villages like La Gaulette will take tourists round the island and out to the two lone rocks behind it. The larger of these is called the Benetiers Rock, a name with which Ile Morne is sometimes confused. The rock rises like a gigantic trumpet coral flower stemming up from the sea. The locals say you can stand twenty-five people on it, and it has one or two small straggly pines growing in its crevices. Its name comes from the locally eaten benetier clam, which in turn is named for its font-like appearance.

The west coast lacks the proliferation of islands of the east, but there is Ilot Fortier in Petite Rivière Noire Bay, and Tonneliers Island in Port Louis harbour, which is now being attached to the mainland by land reclamation. Barkly Island off Fort William was a nineteenth-century newcomer that rose from the sea.

COASTLINE

Mauritius is virtually teardrop-shaped, and with Cap Malheureux as the northern apex of the tear, there is almost no coastline facing due north. This description of the coasts is therefore divided between the east, south and west coasts.

East Coast

This coast is much less built upon than the west coast. The west coast harbours are considered more cyclone-proof, but the reason for fewer weekend *campements* along the east coast is that in winter (June to October) the coast is chilled by 'les brises polaires' from the Antarctic. In this season Mauritians prefer to leave the hills of Curepipe, which they cling to in the hot summer season, and spend their weekends on the west coast in more secluded *campements*, walls or filao screens defending family privacy. Sunday is the day when the beaches are most crowded, but even then the east coast shows long empty stretches.

Cap Malheureux is so named because of the ships wrecked on its reef, though its bay proved sheltered enough to act as the British landing point, near the present chapel, when they invaded the island in 1810. A martello tower stands near the site of the British landing, one of the many the French built along the coast from Cap Malheureux to the south of Port Louis. Their ruins can sometimes still be seen, though many have been incorporated into holiday homes. The coastal road here runs down the east coast through arched flamboyant trees and past pleasant bungalows. At Anse La Raie there is a youth-centre holiday camp for young Mauritians.

The sea at Grand Gaube is somewhat shallow for swimming, but there is a pleasant bungalow-style hotel named after 'Paul et Virginie'. Around the next point is the place where the unfortunate Virginie is supposed to have been washed ashore from the wreck on the Ile d'Ambre, and the Historical Society has erected a little cairn to the memory of the event that Mark Twain described as the 'only one prominent event in the history of the island and that didn't happen'.

Along this coast one may see tobacco growing, though the ground is stony, as exemplified by such village names as Roche Terre (rock land). There is also the contrasting Goodlands. Along this coast, as in many other parts of the island, are *barachois*, little artificial lagoons or sea ponds in which fish are kept or bred. From Grand Gaube the road cuts back inland through St Antoine, a sugar estate where a factory for the production of fibre board for building from the waste *bagasse* or crushed cane stalk was set up in 1971. The road rejoins the coast at Poudre d'Or, named after its golden sands rather than for the supposed treasure from the wreck of the *St Geran* off Ile d'Ambre, which is now in view. At Rivière du Rempart there are eucalyptus groves and sugar mill ruins.

Roche Noires bulges out from the Plaine des Roches district. Across the pale sands flowing 'paths' of black lava rock run towards the cooling sea. In the crannies are rock pools containing small sea fish. It is a slightly melancholy place to visit, the wild ocean and the long fingers of lava reminding one that ships have foundered and men drowned nearby.

The easternmost point is called Poste Lafayette, not after the French revolutionary hero but after one of the men who helped the Abbe de la Caille in the first proper map survey of Mauritius. D'Esny was another helper, and, the Mauritians being thrifty in their choice of names, Pointe d'Esny turns up several times around the coast. There is one nearby in the deep bay on which Poste de Flacq stands, and where there is a headland cairn commemorating seven Special Mobile Force men who were drowned while helping in a sea rescue.

From Flacq the coast southwards possesses some of the most beautiful empty beaches in Mauritius. The Plaines de Belle Mare is the most 'developed'. Here old limekilns have been turned into lookout posts for visitors. Coral is still much used for making lime, being placed in the kilns over a casuarina wood fire and left for a few days. Then water is poured on to produce pure lime. Further south, the beaches of Belle Mare and Palmar may be reached on rough tracks from the main coastal road. The tracks are shaded by filao trees, and terminate in delightful bays of soft coral sand indented with black rocks at the southern end. The beaches are a good source of small but colourful cowries, and the curled bulla shells. There is shade for picnicking at the back of the beaches under the pine trees, and the soft sand slips swiftly into deep reef-sheltered water for swimming.

Below these beaches the coast turns in to Trou d'Eau Douce. Oysters are farmed here, and further south—a new but growing industry. The 'farming' area comprises bamboo frames sunk in concrete in the water, with nets and fertilised eggs. A few miles further south Grand River South East, wide enough at its mouth to need a ferry service, enters the sea, and the coastal plain becomes a narrow shelf between the sea and the mountains of the Bambou range. This part of the coast is especially exposed to cyclones. The people are poor, often living in shack-like houses with palm-thatched roofs.

At Pointe du Diable the coast starts its long westwards swing in to Grand Port Bay. On the Pointe du Diable stand the ruins of French defences, furnished with old guns dating 1750–80. Another reminder that this was once a military zone is the Salle d'Armes rock down on the sea edge near Vieux Grand Port. This is a large flat coral rock, raised above water level, on which the French soldiers would duel either from boredom or over the girls they met in the Bois des Amourettes. The trees of this wood, which still bears that name, were, alas, wrecked in the 1945 cyclone! Swords have been found in the Salle d'Armes area. Seaplanes formerly landed there, and more recently the Russians wanted to establish a fuelling base. Here, too, the first settlements in the island were made at Vieux Grand Port, where a few grass-hidden crumbled walls remain— virtually all the visible relics of the Dutch occupation.

The road then passes through Mahebourg, and joins the main road that cuts right across the island to link Port Louis and Curepipe with the international airport of Plaisance. Below Mahebourg is a wooded peninsula encircled by a road, a popular swimming area. A small hotel/restaurant at one end on the Pointe d'Esny faces out on Grand Port harbour towards the Bambou mountains across the bay. At the other end of a short road through beach bungalows and casuarina pines is Blue Bay, with its shallow, gently sloping beach and views of Ile des Deux Cocos.

South Coast

On the other side of Blue Bay is Le Chaland, where HMS *Mauritius* has a private beach club and there is one of the first beach hotels in Mauritius. The central part of the south coast is rolling sugar-cane country cut by rivers and rising gently to hills behind. Its beaches are not popular—though a few local people have *campements* there—since they lack a protective reef from near Bouchon to Gris Gris beach. Not far west of the end of the reef is 'Le Souffleur', once a splendid spout of sea water produced by waves crashing through eroded coral bores; but erosion has over-enlarged the bores and its performance is now not very spectacular. Even in Pike's time Le Souffleur was losing some of its splendour, though it was mighty enough:

It rises nearly forty feet above the sea, exposed to the full force of the waves and is perforated to its summit by a cavity that communicates with the ocean. When there is a heavy swell the waves rush in and fill up the vacuum with terrific fury. Wave on wave presses on, and there being no other outlet, the water is forced upwards, and forms a magnificent *jet d'eau* ascending to a height of fifty or sixty feet. The noise can be heard for two miles and when the Souffleur growls and roars, it is a sure indication of rough weather.

In 1830 Lt Taylor also described this phenomenon:

A huge mass of rocky matter runs out into the sea from the main land, to which it is joined by a neck of rock not two feet broad. The constant beating of the tremendous swell that rolls in has undermined it in every direction, till it has exactly the appearance of a Gothic building with a number of arches in the centre of the rock which is about thirty five or forty feet above the sea. The water has forced two passages vertically upwards, which are worn as smooth and cylindrical as if cut by a chisel. When a heavy sea rolls in, it of course fills in an instant the hollow caves underneath, and finding no other egress, and being borne in with tremendous violence, it rushes up those chimneys and flies, roaring furiously to a height of fully sixty feet. The moment the wave recedes, the vacuum beneath causes the wind to rush into the two apertures with a loud humming noise, which is heard at a considerable distance.

The main road bypasses the villages of Savannah and Benares, crosses the Rivière des Anguilles and enters Souillac, a large sprawling village in which Mauritius' poet Robert Edward Hart chose to end his days. He is buried in the bleak Souillac cemetery on Cemetery Point, among sailors drowned at sea.

(*above*) Country scene with cane in the foreground and the Trois Mamelles mountains in the background; (*below*) the coloured earths at Chamarel. These are curious ridges of bright and multi-coloured earths found only in this one spot. In the background are the Chamarel mountains

Page 126 (above) Rochester Falls in the south of the island is one of Mauritius' most spectacular waterfalls. The stone has been weathered to the shape of vertical columns; (below) Sambhur deer herd grazing in the Yemen district of Mauritius

In the 1962 Carol cyclone it is said the ground round the cemetery was covered with bones torn from the graves. Also buried in Souillac cemetery is Baron d'Unienville, a historian and the first director of the Mauritius archives. This part of the coast has a harsh look, almost reminiscent of northern Brittany. The savagery of the sea is seen at its most formidable at Gris Gris beach, just east of Souillac. In spite of skull-decorated warning notices of dangerous bathing, however, it is a popular Sunday resort for Mauritians.

Gris Gris means 'witch', and this was also the name of Hart's cat, and mapmaker Abbe de la Caille's dog, after whom the point is supposed to have been named, though some say the name came from the rocks' resemblance to a witch's cauldron. Hart's house, La Nef (the barque), on the road leading east from Gris Gris is now the Robert Edward Hart Memorial Museum. It is a simple house, built by Hart and his friends, and consists of one storey partitioned into a few rooms, with a terrace and steps leading down past a tree screen to a beach strewn with black lava rocks. On the wall by the entrance hang certificates stating that Hart was a fellow of the Royal Society of Arts and licensed by the National Institute of Sciences in London as a practitioner in 'suggestive hypnotism, telepathy, personal magnetism, magnetic healing and suggestive thera-peutics'. The furnishings and personal possessions displayed, such as pith helmet, violin, and paintings, are pathetically few and simple. In an adjoining room is the cot bed where he died alone, and almost destitute, of a heart attack in 1954 when he was sixty-three years old.

Also in Souillac are the Telfair Gardens, named after Charles Telfair, one of the relatively few Britons to come and settle in Mauritius after the British conquest of 1810. He owned the Bel Ombre sugar estate and, as well as doing much to improve and modernise sugar production, was a keen naturalist and a founder of the Royal Society of Arts & Sciences in Mauritius.

Behind the Riambel coastline (*riambel* is Malagasy for 'sunny beach') the green Savanne area is softened with casuarina and Indian almond trees; and on it the Pointe aux Roches is a pile of black lava rocks that attracts local fishermen. This coastline was especially loved by Mauritian poet Paul-Jean Toulet. Backing the rocks in the sand are *veloutier* bushes, with thick velvety leaves. In Jacotet Bay the stones are rolled smooth and round on Mauritius' only pebble beach, so formed by the current of the Rivière des Gallets estuary. In the bay is the raised coral island of Sancho. Bel Ombre, the next village west, looks up to the hills rising to the Black River gorges. Nearby is a cairn to the *Trevessa*, an English ship which foundered in June 1923, 610 miles off Mauritius. A sole lifeboat managed to make an 8-day journey to Mauritius, the survivors disciplining themselves to ration their water till they arrived at the spot where the cairn is placed. In memory of this every year, a charity dinner is given by Mauritians in aid of funds for the Merchant Navy Club. The survival of the *Trevessa*'s crew is recorded in the Mahebourg Museum, where their personal belongings are on show—even to the cigarette tin lid which served as the measure for the daily water ration, and the mouldering remains of the last ship's biscuit ration.

The road climbs up round rocky cliffs before diving deeply to the long canal-like estuary of the River Baie du Cap at Maconde, deep in the hill shelter. It was here that Matthew Flinders was forced to bring his unseaworthy ship in 1803 on its return from exploration in Australia. His joy at having made a safe landfall was turned to frustration and rage when he found the French island at war with his native England, and he was interned. The road then rounds the base of the Chamarel Hills, heading up to Le Morne mountain, which dominates the skyline.

West Coast

The main road cuts across Le Morne peninsula, which has a narrow turn-off to the Morne Hotel. The road, often flooded by the sea in the cyclone season, runs beside the huge peaceful lagoon past the Ile Morne and through little villages intent on fishing and selling shells. On the right are the Chamarel hills and the Black River gorges. North of Petite Rivière Noire Bay are salt pans alongside the road. The villages seen along the way reflect the African origin of the inhabitants. This is the area of sega dancing and the campeachy tree. The coastal uplands even have a faint aura of Africa, with their thorn trees, browned grass plains and herds of deer.

The Baie du Tamarin, Mauritius' surfing spot, is named after the trees which abound in the area. The road now cuts sharply inland, but the next signposted beach is Flic en Flac, one of the most popular on the west coast and crowded with cars at weekends. The fine reef-sheltered sands are backed by a narrow belt of pines and filao. The reef, apart from a small stretch at Petite Rivière, stops here and does not recommence until one reaches Pointe aux Sables. Inland, sugar cane fields rise from the rolling low cliffs up to the base of the Corps de Garde mountain. At Petite Rivière there are sea caves in which hundreds of swallows live, and their nests provide some soup bases for the Chinese community, though few people go there now. The best description of the caves is by Nicholas Pike: 'The Swallows cave [is called] from these birds being supposed to build there in vast numbers; but I could not find one nest. It is about fifty feet deep and eighteen or twenty feet high. It has been formed by the freshlets of the river having washed out the layers of tufa between the beds of lava. The names of numerous visitors were cut in the rocks.'

Though it is impossible to swim hereabouts, the views of the rocky coast from the Caves Point Lighthouse are exhilarating, and the lighthouse keeper will usually allow visitors to see round his small lighthouse, which is situated at the end of a narrow road through sugar cane. Pointe aux Sables is a growing seaside resort among the Indian community, as it is so near Port Louis, and the prominent Mauritian writer Marcel Cabon had his house here. The houses are tucked among thick groups of trees on both sides of the coast road, from which one may occasionally glimpse a narrow sand beach.

Beyond Port Louis comes the deep Tombeau Bay, the estuary of the river of that name; its beach is narrow and built to the edge with *campements* and fishermen's houses. After passing through Arsenal the coast road becomes rough and rutted, but the filao groves of Balaclava, which open out on steep sand-banked empty beaches, are quiet and soothing. The River Citron nearby is choked with water hyacinth. Beyond the Pointe aux Piments the beach is a fashionable *campement* area, and at Trou Aux Biches, considered to be one of the best beaches in the island, a new bungalow-style hotel was opened in 1971. The same good sandy strands continue up to Cannoniers Point. Mon Choisy is a big public beach, and behind it lies a sandy space large enough to act as an air strip for small planes. A plaque commemorates the first flight between here and Reunion in 1934.

Beyond Cannoniers Point, Grand Baie opens out, deep and sheltered, the base for local yacht clubs, with hotels and a restaurant. On the other side of the bay the public beach of Peyrébère is well patronised by Chinese families, with its view of Coin de Mire Island (Gunner's Quoin). The road then passes through private *campements* up towards Cap Malheureux.

10   TOWNS OUTSIDE THE CAPITAL

T HE areas of Vacoas-Phoenix, Curepipe and Moka are linked to Port Louis most of the way by a dual carriageway that curves round the Moka range—a beautiful road coming down through cane fields towards the sparkling sea past central islands of flowering hibiscus and bougainvillea. Before the closure of the railways in 1964, trains fussed their way up to Curepipe; in the early days they were strange double-deckers, first-class downstairs, second-class above. Now the tracks have been asphalted over to make the Sugar Road, a one-way route that flows headlong down into Port Louis among the cane fields to the south of the dual carriageway. In spite of its many crossroads it gives one an exhilarating drive during the off-season, before cane cropping has started.

LE REDUIT

The governor had his special railway coach, now preserved outside the Mahebourg Museum, and a special stop was made at Le Reduit, his residence 6 miles from Port Louis and halfway between Port Louis and Curepipe. Under governors like Sir William Stevenson, who said Government House was in 'the most noisy, dusty and public part of the town', and Sir John Pope Hennessy (1883–9), who preferred the cool calm of the residence to the heat of Port Louis, office staff spent their time rushing back and forth with dispatches and papers.

Even before malaria had driven the inhabitants to the hills, escape from Port Louis had been sought. Governor Bartlemy

131

David (1746–53) built Le Reduit (the refuge) in 1748 at a cost of 80 million livres. Some say it was as a refuge for women and children in case of British attack on Port Louis. Others say it was a hideout for one of his mistresses. Whatever the truth, Le Reduit's creator is commemorated in a stylised little temple at the tip of the gardens built by Governor Sir Hesketh Bell, on which an inscription reads: 'To Bartlemy David, Governor, the creator of Le Reduit, his grateful successors.'

Although Sir Arthur Gordon, governor 1871–4, called it 'our prison between the two ravines', the eighteenth-century house standing on the bluff of a river gorge 900ft up on a spur of the Moka range has a superb setting. The curving drive of camphor trees leads to a beige and white house with pale blue shutters and a black roof. The two wings are joined by a narrow reception room that opens on to a wide terrace over-looking the gardens. The original wooden house was rebuilt in 1778, and the north wing of that house was destroyed in a cyclone in 1868. Inside, the portraits are mainly of British governors, but Labourdonnais gazes sardonically down on the long dining table that seats fifty.

The gardens in the 325 acre estate, laid out by the French botanist Aublet, run along the top of a triangular point between two tributaries of the Grand Rivière North West—La Cascade de Reduit and the Rivière Profonde. The garden contains a pen of giant tortoises and, round the base of a huge banyan tree, a dogs' cemetery. Tip, 'for twelve years Lady Barker's favourite dog', Judy, Fuzzy and Simba lie there. A lily pond and minia-ture falls cut down the middle of the garden, and the point above the two gorges is called the End of the World (Bout du Monde). Below are the cascades of the two rivers, and on the gorge sides thick scrub and trees among which deer roam; in the old days hunts were occasionally arranged at Le Reduit. To the west views over the sea are somewhat marred now by the large white block of the Kennedy College in Beau Bassin.

VACOAS-PHOENIX

Beyond Le Reduit on the dual carriageway the first residential area one reaches on the way to Curepipe comprises Phoenix and Vacoas, which have a combined population of 47,000. This area, with Floreale, east of Vacoas, has a predominantly English flavour. Most of the Diplomatic Corps also live in Floreale, just as Moka is very much the home of the wealthy Franco-Mauritian community. Phoenix is a name well known throughout the island from excellent local beer made at the brewery on the main road. Signposting and well surfaced roads denote a British presence in the area. Apart from Port Louis, only four other communities have town councils of their own— Vacoas-Phoenix, Curepipe, Quatre Bornes and Rose Hill/Beau Bassin.

HMS *Mauritius*

Vacoas is the home of HMS *Mauritius*, a Royal Naval communications station linking stations in the Indian Ocean with London. It was opened in 1961 when the communications station in Ceylon was closed. HMS *Mauritius* is almost a self-contained town, housed in former army barracks of the King's Own African Rifles, which served in Mauritius. It has its own school, church, Naafi, shops, swimming pool and sports grounds. About 250 sailors and officers live in the area, many in the base's own living quarters, as well as about sixty British civilians, including teachers for the base school. About 500 Mauritians are employed and HMS *Mauritius* contributes about £1 million a year to the island's economy.

HMS *Mauritius* exchanges hospitality with visiting ships, picks up daily Reuter news bulletins for the Mauritius Broadcasting Corporation, maintains the broadcasting stations' equipment, and through its social organisations helps in charity work on the island. Its diving club, which has already been

133

mentioned, teaches sailors to dive and has carried out a great deal of underwater exploration. The officers' club is the Naval and Military Gymkhana Club, under the presidency of the captain of HMS *Mauritius*, though there are now a majority of civilian members. Golf, tennis, association football, hockey, squash and swimming are available at the club.

Special Mobile Force

Also housed in the HMS *Mauritius* complex is the headquarters of the Special Mobile Force—the Mauritian army. It was founded in 1960 when the British garrison left the island. The force is made up of volunteers from the police force who do military service for 2–3 years in the SMF, including climbing a mountain a week as part of their training. The men of the force, though technically still policemen, are dressed and trained as soldiers. Its colonel said: 'It's a commander's dream, mostly twenty-three year olds, keen, unmarried men.' The 320 men are mainly employed on anti-riot duties and basic internal security. Under the command of five British regular army officers, who form Britain's loan force in Mauritius, the SMF also forms all ceremonial guards of honour, carries out mountain and sea rescues, finds lost children, dismantles explosives, and helps restore order after cyclones, going over every road checking fallen trees and live electric wires. The men work closely with the police in getting rid of illegal *gandia* (marijuana) plantations and rum stills. Minor riots are settled by the 200 men of the Police Riot Unit, a separate entity, but if bombs are involved the SMF is called in. Breaking up a riot when it involves road blocks can offer problems in Mauritius. The sugar cane fields and most villages have neat piles of lava rocks ready for use in building road blocks and heaving at police. While the SMF is clearing one stretch of road, another road block can be built behind them.

CUREPIPE

Curepipe, on the higher central plateau of Plaines Wilhems, is one of the oldest established communities. Before the malaria plague of 1867 it was just a halt on the main road between Mahebourg and Port Louis. Though probably named after a town in France, its name is popularly thought to come from the fact that soldiers halted there for a breather and to clean out their pipes. In the early nineteenth century it had a small inn whose landlady had a most unsympathetic reputation, and necessity not choice motivated a stay there.

Mauritians say there are two seasons in Curepipe—the rainy season and the season of rains. The area has showers virtually every day, and an annual rainfall of about 125in, roughly the same as London. Visitors at the Park Hotel are often surprised to find it is not raining at the beaches or in other parts of the island.

Finding one's way about in unsignposted Curepipe, which has 51,400 inhabitants and is the second largest community in the island, is made even more difficult by the fact that most residential roads are fenced in by high bamboo hedges, so that each road and house seem identical. Maybe lack of signposting is a defence mechanism against invasion of privacy on such a small island, but it is extremely irritating for the visitor. Some of the oldest and more stately homes of Curepipe are in Lees Road.

The Plaines Wilhems Road, which runs through Curepipe from the dual carriageway, is one of the island's best shopping centres, where Indian and Chinese shops are mixed in with European boutiques. The Royal College, a secondary school for boys, is a stern grey stone building set back from the cross-roads. Near this is the more modern part of Curepipe, with arcades of souvenir shops for tourists and some restaurants. The best of the latter are La Potinière (French) and Tropicana

(Chinese). Nearby is the Casino. Adjoining the casino grounds is Curepipe Town Hall, a wooden two-storey French colonial house that was moved bodily from another part of the island. Its formal gardens contain the inevitable statues of Paul and Virginie, and one of the writer Paul-Jean Toulet. Behind the casino lies the George V football stadium, and facing it the studios of the Mauritius Broadcasting Corporation.

South of the Plaines Wilhems Road stands the Park Hotel, which was once a private house with superb tree fern gardens, but was turned into the island's first hotel for foreign visitors 15 years ago. Sir Winston Churchill Street contains the Botanical Gardens, not so famous or extensive as the Pamplemousses Gardens but still charming; trim lawns enclose a small bandstand, and massed Novembrier lilies wave their dandelion clock mauve or white heads round a small lake. Next to the Botanical Gardens are the offices and nursery of the island's Forestry Department.

The best view of Curepipe is obtained from the Trou aux Cerfs, a 28oft deep and 2ooft wide extinct crater, whence the town and the Trois Mamelles, Rempart and Corps de Garde mountains can be seen. On clear days in May, September and October the Island of Reunion may also be seen from this viewpoint.

ROSE HILL/BEAU BASSIN/QUATRE BORNES

Curepipe is the easternmost residential area from Port Louis. The other three townships—Quatre Bornes, Rose Hill and Beau Bassin—run indistinguishably on from each other, descending westwards towards the sea from the Le Reduit roundabout on the dual carriageway.

Rose Hill shares a council with Beau Bassin, and the two have a combined population of 70,500. Next to its old-fashioned town hall is the theatre and an art gallery founded in 1965. The town clerk of Rose Hill, André Decotter, has become the organiser

136

of the gallery and mentor of young Mauritian painters who wish to exhibit. As well as local paintings, reproductions of European, Australian and Indian paintings are exhibited, particularly for school groups. Up to fifteen or sixteen exhibitions are held each year.

In the Plaza Theatre four to six plays are produced each year, mostly in French. Built in 1929–32 the theatre has a revolving stage and seats 1,300 people, making it the largest theatre in the Indian Ocean area. The French club presenting plays is called La Société des Metteurs en Scène, the British the Mauritius Dramatic Club, and there are other performing groups, as well as visiting singers from Europe and India and ballet and folk dancers. The finals of the annual Youth Dramatic Festival are held in this theatre each year.

Rose Hill's main street has good restaurants and some night life, which is making it an attraction for young Mauritians as well as visitors. Shops, as in Curepipe, stay open later than in Port Louis, till about 5.30 or sometimes 6 pm.

Turning north from Rose Hill's main road, one is soon in pleasing residential areas with elegant spacious houses and mansions such as the Castle, a white turreted building said to be a replica of the Prince of Wales' (Edward VIII's) Belvedere, each room using a different kind of wood; it overlooks a dramatic waterfall, and the Tour Blanche, where Darwin stayed when he visited the island. In this area also are the Balfour Municipal Gardens, small, formal and quiet, with views down into the river gorge and the waterfall and across to the area of Le Reduit. The gardens also possess a pen with some massive giant tortoises.

Beau Bassin, which takes its name from the pool in the Barkly Experimental Agricultural Station, also has the police training school and the island prison nearby, set among peaceful gardens where senior ranking police live. An important medical centre groups the government medical laboratory and the Central School of Nursing.

137

MAURITIUS

Quatre Bornes has a population of 44,915. At Belle Rose there is an open air market every Sunday for fruit and vegetables. Other Sunday markets in Mauritius are held at Vacoas and Triolet (one of the largest villages to the north of Port Louis). These close at 6 pm, and bargains are struck at the last moment by country folk who do not want to take their produce back home.

MAHEBOURG

Mahebourg (pronounced Mayburg) qualifies by its history and not its present run-down state for inclusion in this chapter. At Vieux Grand Port, a huge bay in the south-east guarded by flat reef islands from waves if not wind, the Dutch made their first settlements, and these were continued by the French till Labourdonnais moved the main centre to Port Louis. Attempts are occasionally made to revive the importance of this harbour, which Mauritian historian Auguste Toussaint thinks could be significant in the future. Oil stocks have been built up there, and Russia was recently interested in the Mahebourg area as a refuelling station for her antarctic fleet.

Mahebourg's greatest claim to international historic fame is as the site of the naval battle of Grand Port between the British and the French on 24 August 1810 during the Napoleonic Wars. The British, who had entered the harbour after capturing the Ile de la Passe and lured the French fleet in by hoisting the French flag, drifted on to shoals, and their ships were burned or sunk. Napoleon was therefore able to carve Grand Port's name with pride upon his Arc de Triomphe—the only naval victory there.

The battle is commemorated by a lonely plinth beside the bay, but the real relics of the day are housed in the Mahebourg Naval and Military Museum, the *raison d'être* for most people's visits to the down-at-heel town, lined with small shops where friendliness of service has to substitute for sophistication of

138

goods, and where a purchase worth a few rupees may bring a free drink of coconut milk on a hot day.

The museum, secluded behind the palm avenue, was once the private house of de Robillard, commandant of Grand Port at the time of the battle, and the wounded commanders of the British and French fleets were brought there, in those more courteous days, to convalesce under one roof. Captain ~~Henry~~ Nesbit Willoughby, the English commander, had lost an eye, and Duperre, the French commander, was badly wounded. The two-storey wooden house is now full of historical items. Outside is the governor's private railway coach. Downstairs the early railways, with their curious two-decker coaches, are shown in miniature, and there is a collection of open and closed sedan chairs and litters used before wheeled traffic on the island. Another room contains parts of the ships sunk in the battle of Grand Port recovered from the harbour, including guns from the *Magicienne*, salvaged in 1933, and even pieces of uniform. Prints, maps, the bell supposed to have announced Napoleon's victory at Marengo, a pistol belonging to the local privateer Surcouf, portraits, swords, a collection of Mauritius' stamps, and the bell from the wreck of the *St Geran* are on display. Upstairs there is Labourdonnais' bed, and among the many maps and prints of special interest are the Bradshaw lithographs dedicated to William IV in 1831-2, and a view of Port Louis harbour showing the ranges of mountains visible from the sea, drawn by the draughtsman on board Darwin's *Beagle* on her 1832-6 voyage. Old maps include a Dutch one based on a 1642 sketch by Tasman, in which Port Louis is called Noort Wester Haven and Grand Baie De Boyt Zonder Eyndt (bay without end).

11   SUGAR AND OTHER OCCUPATIONS

SUGAR is Mauritius' staple. Over half her 460,000 acres, or 92 per cent of the cultivated area, is under the waving grey-green plumes of the cane, and 70,000 people work on the estates. Joseph Conrad noted: 'First rate sugar cane is grown there. All the population lives for it and by it. Sugar is their daily bread.' Some people think a one crop economy is disastrous, but as the President of the Mauritius Chamber of Commerce has said: 'We are not growing sugar because we like it. It is imposed on us by the climate. It forms 97 per cent of our exports and we cannot live without sugar.' Although the island and, in particular, the sugar estates are making efforts to diversify, every proposed new crop is faced by two queries—will it survive the cyclones and will it detract from the sugar cane?

Records maintained over 250 years in Mauritius show that one cyclone in every 5 years on average is destructive, yet there is always much of the cane crop left after a cyclone, and even in a bad year growers can expect to save about 60 per cent of it. Great care is taken to produce the best strains of sugar possible. The Chamber of Agriculture was founded in 1853, the first of its kind in the world; and agricultural research was started in 1893, when Philippe Boname created the Station Agronomique, which became a government department in 1913. In 1925 the College of Agriculture was founded, with an emphasis on the factory side of sugar production, and most of the present sugar estate managers were trained at this college. In 1930 the Sugar Cane Research Station was created. The industry could not afford to employ top salaried research staff, and so in 1947 a tax was levied on sugar production to finance the Mauritius Sugar Industry Research Institute (MSIRI), which is now, along with the School of Agriculture, incorporated in the university. The Research Institute has field research stations at Le Reduit (the home of the Institute) in the humid zone, at Pamplemousses in the sub-humid region and at Belle Rive and Union Park. New varieties of cane are grown on the Fuel, Medine and Mon Desert-Alma private estates. Portions of private land are often lent to the Institute for research. The Sugar Cane Research Station and the MSIRI have pioneered work in cane selection and foliar diagnosis, the latter being the analysis of selected sugar cane leaves, instead of soil samples, to determine fertiliser requirements. Quarantine regulations against cane diseases are strictly observed, and every year about 60,000 cane plants are produced. These have usually been selectively bred over a period of some 12 years, and the average cost of developing a new variety of cane for production is about 500,000 rupees.

Sugar was introduced by the Dutch but more for spirit-making than for commerce. In 1639 the first cane was planted in Ferney near Mahebourg. At the end of the seventeenth century Jan Bockleberg made the first sugar on Mauritius. Then Labourdonnais, looking round for crops to make the island more self-sufficient, started growing sugar cane on his own estate, Mon Plaisir, and became a partner in the first sugar factory at Villebague. The ill fated *St Geran* was carrying machinery for the new factory when it was wrecked. Around the island may be seen the ruins of sugar mills and chimneys, predecessors of the modern mechanical factories. The last animal-driven mill was closed in 1853, and in 1862 the last sugar windmill stopped working. The MSIRI Library at Le Reduit contains old prints and paintings of sugar production, and an excellent collection of books on the sugar industry of the world which is available to researchers.

In 1825 the import duties on Mauritian sugar entering Britain were equated with those on West Indian sugar, which gave impetus to sugar growth in Mauritius. In 1868 Dr Icery, the owner of the La Gaiété estate at Flacq, invented the process of sulphuring before liming, and so Mauritius became the pioneer in making white sugar for direct consumption.

Mauritius' biggest sugar purchaser is the United Kingdom, which takes about 380,000 tons a year (out of a total production of about 700,000 tons) under the Commonwealth Sugar Agreement, which remains in force until 1974. The island then hopes to be associated with the Common Market, and to negotiate a special sugar contract with the countries belonging to it. The United States and Canada are also substantial buyers. A year's export earnings on sugar is about 360 million rupees. Rum is also made from the sugar, but its local production was affected by the opening of breweries at Phoenix, though illicit rum has the reputation of having a much better flavour. Certainly the 'official' rum, though fiery, lacks depth of taste.

The sugar estates are the big economic power holders and social leaders in Mauritius, but a surprising percentage of sugar cane is produced by smallholder peasants, mostly of Hindu stock. There are about 27,000 such producers, mostly cultivating less than 5 acres each. They can call on the MSIRI to help them when necessary, and have representatives on the Institute committees. The smallholders have formed themselves into co-operatives to make transporting and selling their crop easier. Even so, 70 per cent of the sugar is owned by companies and private estates, and 62 per cent of the crop is produced on plantations. All the sugar produced is marketed by the Mauritius Sugar Syndicate, and the big Plantation House in Port Louis dominates the skyline, as do the lines of mainly Greek sugar ships that come to collect the crop.

Page *143* (*above*) Beach scene at the Morne Brabant, one of the island's most popular bathing areas; (*below*) an idyllic holiday corner of Mauritius on the south coast where boats may be hired for sailing

Page 144 Entertainments: (above) the Champ de Mars in Port Louis on a race day. Racing takes place here from May to October every year. In the left foreground is the obelisk over Governor Malartic's tomb; (below) the Mauritius Casino in Curepipe. Opened in 1971, it has girl croupiers and a restaurant in addition to two rooms of gaming tables

SUGAR AND TRANSPORT

The sugar estates have had great influence on the transport systems of the island. Originally sugar was often brought round the coast to Port Louis by boat, particularly from such places as Souillac on the south coast. Mules formed an overland method of carrying the sugar. Later came the railways, set up by engineers from Britain in the 1860s. The eventual system comprised lines radiating from Port Louis to the heart of the island. From the main lines the sugar estates ran their own small railways between the rows of cane, tracks which can still be seen today, together with signs that warn of rail cars crossing public roads. The first railway line was the one to the north, opened on 24 May 1864. Before the coming of roads and railways the richer people had progressed round the island in sedan-type litters carried by slaves, accompanied by servants bearing long sticks; later they travelled in wheeled carriages. In all, six railway branch lines were established: Port Louis to Argy in 1864, Argy to Rivière Seche in 1864, Port Louis to Mahebourg in 1865, Rose Belle to Souillac in 1878, Rose Hill to Saint Pierre in 1880, and Richelieu to Tamarin in 1904.

Though the sugar industry was a heavy rail freight user in the cropping period from June to December, the railways proved uneconomic and in 1964 they were finally closed. Before this, agreement had been reached between the government and sugar estates that the latter would help finance the new roads. This was done, and the country now has a remarkably extensive complex of roads, with more building or under improvement,

A devotion to their main crop has given Mauritians a sweet tooth. Every toddler sucks sugar cane as he follows his mother working in the fields. In town small children and street hawkers sell sweets in glittering mounds. In offices, when tea is offered,

I

it always comes with milk and plentiful sugar added, and you are looked at askance if you say you do not take it.

SUGAR WELFARE SCHEMES

The sugar workers live on the estates, and the owners finance a Sugar Industry Labour Welfare Fund Committee to look after their employees. The fund's local centres are easily recognisable in villages by the initials SILWF painted on gateposts. The centres act as dispensaries and give family planning aid. The fund is mainly spent on building housing estates in villages where the labour force resides, providing the welfare centres and financing scholarships for workers' children. Although the planting is done in March and harvesting from June onwards, there is plenty of in-between work in weeding, fertilising and combating crop diseases and pests. Each worker is estimated to cultivate about $3\frac{1}{2}$ arpents ($3\cdot64$ acres) of cane.

Mauritius' population has nearly doubled since the end of World War II, and is expected to top a million in the 1970s, meaning that more jobs must be found and that more food must be produced to feed her people and reduce imports. The sugar plant itself is fully used. The sugar estates burn the *bagasse*, or crushed cane stalks, for fuel, and the surplus is now being used to produce fibre board for building purposes.

INTERCROPPING

Each row of cane is planted about 5ft apart, which means that there is a great opportunity for intercropping. Sunflowers and maize were tried, but grew too quickly and competed with the cane for light and moisture. The chosen intercrop must grow at the same pace as the sugar, be planted at the same time, and not in any way detract from the sugar yield. The crops that have so far passed this test are potatoes, tobacco, beans and tomatoes (*pommes d'amour*, grown in the northern more rocky soils). In 1972 about 10,000 tons of potatoes were produced. Now seed potatoes are also being developed, since the cost of importing seed amounted to half the total production costs.

OTHER CROPS

Mauritius is mainly a rice-eating community, spending 60 million rupees on rice imports each year. It also needs food for its growing tourist industry. It is hoped to produce vegetables and fruit such as pineapples and some citrus to feed the hotels. Onions are grown on many small holdings, melons on the Riche-en-Eau estate in the south-east, mangoes and pawpaws abound in the trees, and bananas grow easily, though are often destroyed in the cyclones. Groundnuts have proved an easy grower, and now ginger is being developed as a commercial crop. Another crop that has proved successful is aloe fibre, produced from the furcraea gigantea plant, also known as 'Mauritius hemp'. In 1963 1,300 tons were grown. In 1971 1,976 tons were produced from which 2,599,000 sacks were made. The crop is baled, stored and marketed by the Mauritius Hemp Producers' Syndicate, the whole output being sold to the Government Sack Factory. Most of the sacks made from it go to the local sugar industry, and the rest are exported.

Tea

Mauritius' second crop has become tea, grown commercially since 1886. With its large labour force, the ability to grow on land where sugar cannot grow, and reasonable resistance to cyclones, the crop seems ideal for Mauritius. Areas of indigenous forest on the central plateau are being cleared to grow it and several large tea factories have been set up.

Indian types of tea are principally grown and they earn about 14 million rupees per annum in exports to Britain and increasingly to South Africa. As with freeze drying of vegetables and other diversification projects, the sugar estates have invested heavily in the production of tea. Coffee interests them less because of its lack of resistance to cyclones, the plant being in flower during the cyclone season. The coffee grown in the Chamarel mountain area is sheltered between hill ridges.

Tobacco

Another important crop is tobacco, with about 500 planters producing about 600,000 tons a year. The Dutch are credited with first introducing tobacco to the island, but various taxes later discouraged its expansion. In 1926 the British American Tobacco Company established itself in Mauritius, and since 1930 there has been a Tobacco Board and government warehouse controlling production. About 1,000 acres are under tobacco, supplying three cigarette factories, which blend it with imported leaves. The three companies producing cigarettes on the island are British American Tobacco, Consolidated Tobacco and Amalgamated Tobacco.

CATTLE, PIGS AND FISH

In addition to new intercrops, the sugar estates are looking at various types of livestock breeding to cut Mauritius' large imports of poor meat from Madagascar. (Home-killed meat is often made tougher by the Muslim slaughterers' bleeding-to-death methods.) There is certainly no lack of cattle reared on the island. We have mentioned the 30,000 cows owned by smallholders and kept for milk. The calves of these animals are usually slaughtered when they weigh 75–100 kilos, and it is argued that this is wasteful, since Mauritius is rich enough in feedstuffs to raise the animals to 400 kilos before they are killed. Molasses, another by-product of sugar, is being used for animal feeds on the island and experts feel with the use of fertilisers, hormones and other aids, the cattle industry could help Mauritius become more independent in feeding herself.

One of the leading cattle raisers in the island and a former president of the Mauritius Stock-Breeders' Association is Alain Cambier of Le Morne estate. He now has a herd of 1,000 cattle and 800 deer on his 2,500 acres. He believes cattle can be reared on molasses and *bagasse* (with imported artificial protein added), allowing the £12 million rupees spent on Madagascar beef annually to remain in Mauritius. The Mauritius Stock-Breeders Association consists of the twenty-one ranchers with the biggest herds, mostly raised on sugar estates like Fuel, Medine, Savannah, Constance, St Felix and Mon Choisy. Cambier is experimenting with artificial insemination on five different cattle breeds over a 2 year period to determine which strain thrives best under Mauritian conditions.

More pigs also could be bred in Mauritius. The creoles love them, and most creole village families keep a pig or two in their gardens along with chickens. Pigs are imported from the dependent island of Rodrigues and there is talk of producing hams from them in Mauritius. But, as with cattle, there are religious difficulties preventing increased consumption of pig meat throughout the island; the Chinese and creoles would accept it but the Muslims not. Venison suffers from no such taboo, and some experts say 15,000 deer could be bred on land where nothing useful now grows.

Culling the harvest from the sea has only just got under way. We have seen that oysters are being farmed, and attempts are being made at farming camaron prawns. Japanese experts came in 1972 to study the possibilities of breeding other varieties of prawns commercially.

There are 95 square miles of lagoon for fishing as well as 200 square miles of sea shelf to the north of the island. The tuna fishing fleet of Japan, based in Port Louis since 1963, fishes beyond the reef. The Department of Fisheries has an inspectorate to ensure that fishing is carried out legally among the 3,000 fishermen, but dynamiting fish still occurs. A small fish farm experiments on farming the freshwater tilapia, which is rich in protein.

149

The interaction of cultures and the early necessity to survive in an island of limited resources should have led to a richness in local arts and crafts. But this has not happened, and while the poorer women are superbly self-sufficient family providers, they have not the time or, as yet, the markets to produce items to be sold primarily as tourist souvenirs. With a growing number of visitors, however, the government has been concerned to develop more handicrafts. A small boutique for Mauritius handicrafts has been set up in the Company Gardens at Port Louis, and craftwork is on sale there. A further kiosk is being opened in Pamplemousses Gardens. These mainly display the *tente* baskets of all shapes and sizes (yet more baskets can be seen in the Port Louis market) and the same type of woven work made into table mats and little slippers. Sugar cane pictures are attractive if more rarely made now, and there are shells and items made from goatskin.

Of all crafts, Mauritians are most proficient in needlework. Indian girls have a natural talent for it, and, at the other end of society, the influence of French couture has created a high standard of dressmaking. The wealthier women, continually seen at close quarters by the same small island society, spend a great deal of time at their dressmakers. Madame Henry Lee is the acknowledged doyenne of the business, and has her own boutique in Curepipe. Materials are bought at reasonable prices in the Indian fabric shops in the town and made up to requirements.

The government has set up training centres to teach sewing to village girls in country areas. (For 10 years boys have been taught metalwork and woodwork.) Twenty girls are, for example, being trained at the government centres in Triolet and Surinam villages. But once the course is finished, the girls, though they sew for their families, have not been using their talents for the good of the community. Because of this a group of leading Mauritian women, predominantly Indian, set up the Women's Self Help Association. They built up a fund for their work by organising charity events, so they could buy materials and designs for girls in villages to work on and embroider. The work has now prospered so well that the Association has opened a non-profit-making shop in Rose Hill, and is now getting orders from abroad. The work produced includes tablecloths and household linens, cross-stitched rugs, traditional kaftan-style dresses, children's clothes and superb modern nightwear in cool cotton fabrics.

The Association, which cannot be termed 'women's lib' in any way, except so far as it is trying to help the poorer women to help themselves, consists of fifty members who meet once a month, with a twelve-member committee. Each committee member is in charge of a group of village workers. The work has gone so well that village girls are able to earn up to 75 rupees a month while remaining in their own homes, an important consideration in Hindu communities, where the men do not like their women being exposed to the public gaze by, for example, serving in a shop. The Association plans to set up its own trading centres, in particular in the neglected corners of the island. Leather sandals, cane and basket work were also produced for the Association by boys, but the production flow was found not to be so reliable. There is now a leather crafts factory near the airport.

Each garment produced by the Association is labelled specially, and fashion shows and exhibitions of the work have been given to raise more money for training. It is hoped to open more shops in different towns, and the Association aims for ten teaching centres and 100 girls each year, of which half will remain permanent workers.

The same meticulous handwork which goes into these garments can be seen in the reproduction of model ships at the Jose Ramart factory in Curepipe. Here exact scale models are produced of famous historic ships such as Nelson's *Victory*, Darwin's *Beagle*, the *Mayflower* and various French vessels. Everything is done by hand, from making the little bronze cannon to the complicated rigging. The sails are hand-stitched and flags embroidered by hand on both sides. The sixty-four workers can produce one ship a day, and the largest costs about 880 rupees sold direct for export. The same firm also reproduces naval furniture—brass-bound chests, writing sets in wooden boxes and ship's travelling bar sets.

Antiques and reproduction furniture are widely sold in Mauritius, the latter beautifully produced by local craftsmen. Painting is another art form growing in strength with the encouragement of the art gallery exhibitions in Rose Hill. Malcolm de Chazal has already been mentioned. Other leading artists include Max Boullé, Serge Constantin and Hervé Masson. Sculpture arouses interest, though few exhibitions are held. Mauritius' best known sculptor was Prosper d'Epinay, whose statues of royalty and officials can be seen in Port Louis.

STAMPING GROUND

One item Mauritius has become internationally famed for is her stamps. The Mauritius penny red and twopenny blue issued in September 1847 are among the most valuable in the world. The 1972 Stanley Gibbons catalogue put the value of each at £22,000 (£15,000 used), and £30,000 was paid for a 'blue Mauritius' in Frankfurt in 1972.

Mauritius was the first British colony to issue adhesive stamps, and in 1847 Joseph Barnard, a Port Louis engraver and jeweller, was given the job of engraving the copper plates for them. He became confused and tried to check with the post office. It was closed, but seeing the words 'Post Office' on the doors, he engraved this instead of 'post paid' on the side of the stamps. Lady Gomm, the governor's wife, had ordered 1,000 stamps of the issue in advance, as she wanted to use the new issue immediately to post invitations for a fancy dress ball. It was therefore decided to issue the stamps and the whole set was released; subsequent issues carried the correct 'post paid' words and even these, issued in 1848, are now worth about £4,000. Only twelve of the 1847 stamps are known to be still in existence.

Since then Mauritian stamps, which are printed in London, have always been popular with stamp collectors, and the post office makes a profit partly from being able to sell about 50 per cent of each issue to collectors (compared to about 25 per cent in Britain). Issues have shown shells, fishes, marine life, scenery, and aspects of the sugar industry. Most recently they have commemorated the visit of Queen Elizabeth in 1972, the 150th anniversary of the Port Louis Theatre, and 'pirates and privateers'.

12 PROSPECTS FOR THE ISLAND

THE difficulties of trying to please all sections of Mauritius' mixed community delay decisions. The island is French by culture, a member of the British Commonwealth, receiving aid and technical help from India and money from China, and eyed by Russia as a useful Indian Ocean base for the Soviet Antarctic fleet. She is geographically nearest to Africa, though her Prime Minister, Sir Seewoosagur Ramgoolam, points out a natural attraction to India, since that country's recent economic problems were similar to those now being experienced in Mauritius.

In addition to all these influences, Mauritius is strongly attracted to Europe, having contracted to associate herself with the Common Market countries under the Yaounde Convention, which governs an association of eighteen African countries. Mauritius will lift trade barriers in return for community aid of up to £2 million by the beginning of 1975, plus grants from the European Development Fund.

Population growth and lack of jobs are the greatest problem facing Mauritius. A 4 year development programme has been drawn up by the government to create some 130,000 new jobs, and the family planning programme has reduced the population increase to 1·6 per cent per annum. But labour is Mauritius' only cheap natural resource, and with this she is hoping to attract many new industries into the free zones she created in 1970.

Plans for future employment now that sugar, with mechanisation, seems to have reached the zenith of its labour absorption

154

are directed towards (1) relief work for the unemployed, (2) the opening of free zones for new industries, and (3) the development of tourism. Emigration to countries like Australia also continues. Even six men emigrating makes headline news in the local press, such is the urge to get out and get going among the young. Twenty-five nurses probably chosen from hundreds of applicants go to Britain—again front-page news with photos. Ministers' offices are surrounded by a pressing sullen throng begging for jobs and chances. The government considers such soliciting a relic of colonialism, unnecessary now that there is a Public Service Commission.

THE DEVELOPMENT WORKS CORPORATION

The Development Works Corporation is a government department that was set up in July 1971 to rationalise relief work and to create more jobs. It is meant to be temporary. In 1972 it ran ninety-five work sites employing over 5,000 persons on building roads, schools, post offices, and village halls, and on afforestation, mixed farming projects and increasing livestock production. Groups of workers are paid 40 per cent in kind, with food such as rice and milk, and 60 per cent in cash; and they work 6 days a week from 7 am to 3 pm. The wages paid are deliberately slightly below average in order not to discourage workers from taking permanent jobs. The work does not include maintenance or cleaning projects, but concentrates on projects that will in turn create more jobs.

One of the biggest schemes is in the Midlands area, where 12,000 acres are being cleared for mixed farming, and 600 families have been settled. On the coast *barachois* fish ponds are created and protective walls and sluice gates built. New coast roads are being built, giving better access to the beaches. Forests, in particular wood for the match industry, are being planted on bare mountain slopes. In Rodrigues it is planned to build a school and roads. Irrigation is needed, especially in the north of Mauritius, and more pipes need to be laid for drinking-water supplies.

FREE ZONE INDUSTRIES

Mauritius' industry has been largely aimed at making her more self-sufficient. She makes her own matches and candles, repairs all her own buses and now exports bus bodies to Reunion and Madagascar. Furniture and louvre windows are also made. While Mauritius has nothing but her sugar and tea to export in any substantial quantities, and over a long distance with heavy freight charges, she has wisely decided to concentrate her future on light industry, playing up her cheap labour as a lure for investors. In Mauritius the semi-skilled woman earns 2·50 rupees a day compared with around 20 in Hong Kong, and a skilled woman will earn 4 rupees compared to a man's 7.

Free zones (known as Mauritius Export Processing Zones) for factories have been created, the first being Plaine Lauzun just outside Port Louis. At Coromandel, a charming site across the Grand River North West, development is already under way. Investors are offered 10–20 year tax holidays and free importation of machinery provided a minimum of 20 local people are employed. So far the main countries investing have been South Africa, Hong Kong, Madagascar and Germany, and France and Britain have shown interest. Britain has established an electronic capacitor factory. The majority of the items made are for direct export, and in most cases are expressly aimed at using labour. 'We export holes', said the director of one firm, which does nothing but drill holes in components for watches and then re-export them. Another company developing well imports diamonds from London and returns them cut and polished, with a great saving in labour costs.

Textile making-up is another growing branch of industry, consonant with the Mauritians' natural talent for neat and quick handiwork. Shirts were already made locally and new glove, knitwear and clothing factories are being opened up. A Chinese-German sponsored soft toy factory has been started with exports exclusively to Germany, and there are several wig factories. When Mauritius' first hotel was opened in 1952 all the furniture was imported; now it is all made locally. One firm exports carved furniture legs to South Africa. Edible oils are made into margarines, and the Mauri-foods company is engaged in producing quick-frozen vegetables and fresh products for local hotels.

Sea freight is invariably costed to include an empty leg return journey and is therefore deterringly high, and recent dock strikes have not encouraged shipping lines. The new industries, therefore, concentrate on lightweight items that can be sent by air freight. Part of Mauritius' future development involves the construction of a new international airport, which will take 4 years to build and cost about 50 million rupees, mostly in loans. It will take jumbo jets and reduce the present difficulties of crosswinds and night landings. Two plans are under consideration—an extension to the present Plaisance site in the south near Mahebourg, or a new airport in the flatter northern plains near Pamplemousses, which would be easier to land on.

TOURISM

Tourism is often a mixed blessing in any economy, yet the prospects are it could overtake sugar as Mauritius' major source of revenue. It already ranks second. Tourism began in 1952, when the first flights from Australia to South Africa landed in Mauritius. The airline agents were asked to find accommodation for 50 passengers and crew overnight. Though time was short, they bought a graceful private house with superb gardens in Curepipe, which became the Park Hotel. Next, a small rondavel beach-hut hotel was set up where the Le Morne Hotel now stands, and another hotel was set up at Le

Chaland near the airport in 1962, when the BOAC services started.

Now Mauritius is well served by most of the major international lines. Air France, in pool with Air Mauritius, runs frequent link services with Reunion and Madagascar. Reunion has been Mauritius' greatest source of tourists, many coming with their families for the long school-vacation periods. South Africa is another good source of visitors, as was Rhodesia, but European visitors are increasing, with talk of charter planes coming soon. Links between the Seychelles and Mauritius have been established by BOAC (now British Airways).

Earnings from tourism rose from 15 million rupees in 1966 to 24 million in 1970, when 27,650 people visited the island. In 1970–1 the tourists spent 35 million rupees.

Tourism must be integrated in the life of the island, says the government. Yet beaches must not be taken away from the people. The New Mauritius Hotels group has plans, however, to make the whole of the coastal area from Tamarin and the Black River area round to Maconde into a giant 'micro-urbanisation' aimed entirely at tourism. Other plans will develop some of the offshore islands, and the north-west coast.

Hotels

To date the hotel scene has been dominated by the New Mauritius Hotels group, which has provided the following good-class hotels for visitors: Le Chaland; Le Morne Brabant, now being enlarged with new room blocks; the Trou Aux Biches, a beach-bungalow complex opened at the end of 1971 15min north of Port Louis; and the Park Hotel in Curepipe, used mainly by businessmen. The latter are now beginning to use Trou Aux Biches, which is nearer Port Louis. The New Mauritius Hotels group has its own dance band, which tours its establishments.

The Ambassador Hotel was opened in late 1971 in Port Louis for a world health conference. It is sited close to the Parliament and City Hall buildings on the corner of Sir William Newton and Desforges streets. The rooms, though simple in decor, are air-conditioned, and have private bathrooms. There are three smaller hotels of tourist standards in Port Louis, and a new 150 room hotel is planned. The Sunray Hotel at Coromandel, near the industrial free zone site, was opened in 1972.

In the past few years a number of small boarding houses and privately run hotels have come into existence, though they cater more for Mauritians than international visitors. Well established are Blue Bay Hotel on Pointe d'Esny in the south, and Touessrok on its tiny islet off Trou d'Eau Douce. In the north-east at Grand Gaube is Paul et Virginie Hotel, and at Grand Baie the Ile de France Hotel, with swimming pool, good local food and, on the opposite side of the road, a small discotheque.

Recently there has been a tendency to set up small villa and bungalow complexes (Paul et Virginie has self-catering kitchens in its bungalows) for the use of long-staying families, mainly from Reunion. Le Morne has twenty-four bungalows with kitchenettes set along the beach, and places like Villas Caroline at Flic en Flac, Etoile de Mer at Trou Aux Biches, Kuxville at Cap Malheureux and Merville at Grand Baie fulfil this need. Visitors can rent private *campements*, but these are usually booked a long way ahead by families from Reunion, who find the price of living much cheaper in Mauritius.

British Airways has shares in New Mauritius Hotels and has developed through another company (Dinarobin Inns and Motels Ltd) Trou Aux Biches and the extension at Le Morne. Club Mediterranée, the French holiday company, is planning a hotel at Pointe aux Cannoniers in the north-west, and there are plans for German and South African investment in hotel development. Including the hotel planned by the Club Mediterranée, five new hotel projects were expected to be under way by 1973, all financed by foreign capital.

To date there are 716 good standard rooms (1,432 tourist

beds) in the island. With the projected new hotels it is estimated that there will be 1,356 rooms (2,712 beds) by 1973. The number needed by 1975 is 2,018 rooms (4,036 beds).

Sightseeing Tours

Mauritius is not well organised for tourism, and its amenities cater more for the individual who wants to explore quietly than for large groups. However, excellent day and half-day tours are arranged from the main hotels by the Mauritius Travel and Tourist Bureau in Port Louis, using mini-bus transport. These are reasonably priced at about 8 rupees for half a day and 16–22 rupees for a full day. The various tours cover all the main coastal beauty spots—Pamplemousses, Port Louis, Mahebourg. Tours also take visitors to see the deer in the Yemen game reserve, a private estate.

What to Wear

Clothes should be light all the year round on the coastal areas, but visitors to the central plateau or mountain areas also need some woollens and a raincoat. Waterproof garments in the cyclone season are necessary in all parts of the island, and some light waterproof boots or rubber sandals would be helpful. Though malaria has been eradicated, there are still possibilities of insect bites, and insect repellent and healing lotions should be taken to beach hotels, most of which have air-conditioned rooms.

Information Sources for Tourists

Mauritius Government Tourist Office, La Chaussée Street, Port Louis; and Mauritius Travel and Tourist Bureau, Head Office, corner Sir William Newton Street and Royal Roads, Port Louis.

13   MAURITIAN MISCELLANY

THE most important item in any island's life is its communications, especially when that island is off the main air routes. Mauritius, since the decline of shipping calling at Port Louis for fuel, has developed good air links, and her postal services to Europe only take two days.

AIRLINES

Mauritius' international airport, Plaisance, is situated inevitably among cane fields at the south-east of the island near Mahebourg. It is 27 miles from Port Louis, 28 from Le Morne and 5 miles from Chaland. Pilots are said to dislike it because of tricky crosswinds.

Major lines, eg Air France, British Airways, Qantas, East African Airways, Air India, Alitalia, Lufthansa and Zambian Airways operate through Mauritius. Air France provides almost a daily service to Paris, with links to London. Air Mauritius, which makes flights to Reunion, will eventually serve Rodrigues and London, and has already pool agreements with South African Airways. There are plans to link Mauritius with the Seychelles, so that tourists can take two-centre holidays. The airlines are represented in the main by Rogers & Co in Port Louis, the island's leading agent for shipping and airlines, though East African Airways and Lufthansa have separate representation.

MAURITIUS

There are now no regular passenger ships calling at Port Louis, though the Clan Line still runs ships from Britain via South Africa to the island. Occasionally large passenger ships, often belonging to the Union Castle or P. & O. lines, visit Port Louis, mostly on cruises from South Africa. Every month MV *Mauritius* makes its voyage to Rodrigues and the Seychelles.

ENTRY REGULATIONS

British and Commonwealth citizens require only a valid passport. Other countries not requiring a special visa are Denmark, Finland, France, Greece, Iceland, Italy, Lichtenstein, the Netherlands, Norway, San Marino, South Africa, Spain, Sweden, Tunisia, Turkey and Uruguay. Visas are only required for Germans and Israelis if they stay more than 3 months. Citizens of other countries require a visa. Certificates of vaccination against smallpox must be produced by all visitors ,and those entering from cholera or yellow fever areas must produce certificates of inoculation. There are no currency import restrictions, and the airport has banks for exchange. A maximum of 700 rupees in Mauritian notes can be imported and 350 taken out.

CUSTOMS REGULATIONS

Personal belongings, 250 grams tobacco, 25 centilitres of toilet water, 2 litres of wine, ale or beer and 75 centilitres of spirit can be brought into the country without customs duty, and up to 1,000 rupees worth of souvenirs taken out without an export licence. Firearms and ammunition must be declared on arrival. A free listener's licence for 2 months is granted to visitors bringing radio sets. No sugar cane may be brought into the island, and other plants only with a previously obtained plant import permit, obtainable from the Department of Agriculture. The Department also issues advance entry permits for live animals, and all animals have to go through 6 months' quarantine on arrival.

CURRENCY

The unit of currency is the rupee, divided into 100 cents. At the time of writing the exchange rate is 13·34 rupees per £1 sterling. In 1971 new coinage was issued: the 200 rupees coin shows 'Paul et Virginie', the 10 rupees the dodo, the 1 rupee the coat of arms, and the half rupee piece a stag. Banks are open from 10 am to 2 pm on weekdays, and 9.30–11.30 am on Saturdays.

ROAD COMMUNICATIONS

The system of over 600 miles of surfaced roads is good, much better than mapping, signposting and road-naming. Though there is a good dual carriageway linking Phoenix-Vacoas with Port Louis, other major road developments are being held back until a decision has been taken on where the new airport will be sited. A swift road linking Plaisance airport with Port Louis is badly needed, as is a northern road linked with a bypass or ring road round Port Louis; at present all through traffic has to travel via the city centre. Vehicles are driven on the left, and visitors hiring a self-drive car from firms such as Avis or Mautourco (Mauritius Touring Co Ltd) can get help from the Automobile Association of Mauritius, 2 Queen Street, Port Louis. A valid driving licence is necessary before one may hire a car. The average driving speeds around the island are 30–40 mph.

Taxis, recognisable by white number plates, are plentiful and reasonably priced, but the return mileage is usually charged and fares should be settled before starting a long journey. They have no meters. There is a fixed rate from most of the hotels to the airport. Tips are not normally expected by taxi drivers or hairdressers. Bus communications round the island are extremely good and very cheap, each major community's service being linked with others. Visitors are using the buses more and more, and a list of routes is included in the Mauritius *A–Z Tourist Guide*, which also includes vital street maps.

TELEPHONES AND POST

The telephones are run by the Telecommunications Department. External links are by sea cable and radio and run by Cable and Wireless Ltd. There are twenty-five telephone exchanges in Mauritius, including rural automatic exchanges. Individual subscribers numbered 19,113 in 1971.

There has been an international telex system since 1969. A permanent radio telegraph station at Vacoas maintains a twenty-four-hour watch on the international maritime distress frequency, and provides communication for shipping in an 800 mile radius. Meteorological broadcasts are received here and communication is made with meteorological stations in the dependent islands.

Parcels are usually sent seamail and all letters abroad go by air. Within the island there are two deliveries a day. All mail is sorted at Port Louis and then sent round the island in the Post Office's own vans. There are eighty-five post offices around the island.

TIME

Standard time in Mauritius is 3 hours in advance of British Standard Time and Mid-European Time.

ELECTRICITY

Mauritius' electrical supplies come from a number of hydro-electric plants, such as the reservoirs Mare aux Vacaos, Mare Longue, La Ferme, Eau Bleue and Tamarin Falls.

WEIGHTS AND MEASURES

Miles are indicated on signposts but the metric system is still mainly used. The most common land measurements are a French foot (1·06 English ft), an *arpent* (1·04 acres), and a *toise* or 6 French feet (2yd 4in). Other unusual measurements, of capacity, include a *bouteille*, which equals 800cc (liquid); a *chopine*, or half a *bouteille*; and a *corde*, used to measure firewood and equivalent to 96·82 English cu ft. In the sugar cane fields, the measure of length mostly used is a *gaulette*, which equals 10 French feet. The term *livre* means half a kilogram.

14 THE DEPENDENT ISLANDS

THE first catalogue of the dependencies of Mauritius, made in 1826, unfortunately included sandbanks and some imaginary islands; but since 1921 the following group has come under Mauritian governorship—Agalega, the Cargados Carajos, the Chagos Archipelago and Rodrigues.

Agalega, 580 miles north-west, comprises two islands separated by a sandbank; its 400 inhabitants produce copra and coconut oil for the margarine industry of Mauritius. The Cargados Carajos, a group of twenty-two islands 250 miles north-east, are usually referred to by the name of the main island—St Brandon. These islands are a fishing base leased to a Mauritian company, and dried fish is sent to Mauritius from St Brandon. Diego Garcia in the Chagos group 1,180 miles away, was once Mauritian, but in 1965 was ceded to the British Government for 40 million rupees. It is proposed to set up a joint British-American naval tracking and air base there.

The Chagos group, Agalega, and the Cargados Carajos lie in the path of trade winds and cyclones, the latter averaging about nine a year. Their total area is about 47½sq miles. The islands are coral atolls mostly covered with coconut palms, and because of the coconut oil produced they were once known as the Oil Islands. Some of them are hardly more than sandbanks, with some small scrub vegetation. Coconuts, fishing and the collecting of guano are the main activities. Though each community is dependent for welfare on the company to which the right of developing the resources belongs, they are admini-

166

stered by a visiting magistrate. Teachers, technicians and nurses are made available by the Mauritius government.

Rodrigues is the most important dependency, 360 miles east. It was discovered in the 1530s either by the Portuguese explorer Fernandez Pereira or by Diego Rodrigues, after whom it was named. The first settlers were a band of Huguenots under François Leguat in 1691. They left in 1693 for Mauritius with their pieces of precious ambergris, which led to their imprisonment for two years on the Ile aux Vaquois by the Dutch. Rodrigues remained uninhabited till 1725, when it was occupied by the French, but most had left by 1803. In 1809 the island was taken by the British, and became the rendezvous of the British forces, 16,000 sailors and soldiers, which moved on to capture Mauritius. Under the Treaty of Paris in 1814 Rodrigues was ceded to the British crown with Mauritius. In 1846 the first police magistrate was appointed in Rodrigues.

Today the island has about 26,000 inhabitants, mostly farmers and fishermen. Life is harsh, bedevilled by cyclones that destroy crops, animals and boats—the props of livelihood. In 1968 the cyclone Monique was reported to have produced winds blowing at over 170 mph.

The 40sq mile island, surrounded by a coral reef, is a fish-shaped volcanic sea mountain older than Mauritius; its central spine, running at 2,000ft, has north/south spurs and deep ravines. The island is dry, with barren slopes on which sheep and goats graze. Most of the population are scattered on small-holdings in the hills, for which they pay a nominal rent to the government. Maize is the main crop, and poultry, pigs, beans, onions, garlic, lemons and salted fish are exported to Mauritius. Rodrigues could become a useful supplier of meat and vegetables to Mauritius.

The island remains remote and uncomfortable to visit. MV *Mauritius* visits it once a month to bring supplies and officials working on the island, and take away its exports. A hotel of 25–40 rooms is being planned, and a small air strip was opened in 1973. The only other connection with the outside world is the 6-monthly visit of the fishing boat from the Raphael Fishing Co, which hires men on 6-month contracts to fish at St Brandon.

There is no cinema, TV or newspapers, radio reception is poor and the electricity supply is limited to Port Mathurin—so it is not surprising that there is no rush on the part of officials to go and live there. The 1968 Report on Rodrigues expresses the isolation of this island: 'No dental surgeon unfortunately visited the island during the year . . . Likewise no eye specialist came and there are numerous cases which require immediate cure.' It is hardly surprising, considering the lack of communications, that the Report notes that 'no prisoner escaped during the year'.

Rodrigues is administered from Mauritius by a magistrate who does not have to be resident, and a civil commissioner who co-ordinates the various government departments. Four parish councils work with him. There is a court on Rodrigues, with appeal to the Mauritius Supreme Court. The law is the same as in Mauritius; there is a commissioner of police, and a police force enlisted on a monthly basis and trained in Mauritius. There are seven primary and two government schools; one of the latter is at Port Mathurin, the only real community (about 15,000 inhabitants), and one at Oyster Bay. There are several Roman Catholic schools, and in Port Mathurin two secondary schools—one Roman Catholic and the other Anglican. Though most of the population, descended from French and African stock, are Roman Catholic, the Church of England has a priest and church on the island. There are also a few Hindus and Chinese living there.

BIBLIOGRAPHY

ADMINISTRATION, COMMERCE AND INDUSTRY

BUCKHORY, S. (ed). *Port Louis Handbook of the City Council* (Port Louis, 1966)
——. *Our Constitution* (Port Louis, 1971)
MEADE, E. *The Economic and Social Structure of Mauritius*, Sessional paper No 7 (Port Louis, 1960)
SWINDEN, J. B. *Local Government in Mauritius* (Port Louis, 1947)

AGRICULTURE AND GEOLOGY

AYRES, P. B. 'On the Geology of Flat and Gabriel Islands', *Royal Society of Arts and Sciences of Mauritius Proc*, Part II, Vol II (1865)
CLARK, G. 'Notes on the Geological Features of Mauritius', *Jnl Geol Soc* (1867)
DE HAGA, HAGA, H. 'Physical Features and Geology of Mauritius', *Jnl Geol Soc* (1895)
KELLER, C. *Madagascar, Mauritius and Other East African Islands*, trans H. A. Nesbitt (1901)
MAURITIAN SUGAR INDUSTRY, PR OFFICE. *The Sugar Industry in Mauritius* (1972)
NORTH COOMBES, A. 'Tea in Mauritius 1817–1944', *Revue Agricole de l'Ile Maurice*, Vol 23, No 6 (Nov–Dec 1944)
——. *The Fibre Industry of Mauritius* (Port Louis, 1951)
SHAND, S. J. 'The Lavas of Mauritius', *Jnl Geol Soc*, Vol 89 (1933)
SIMPSON, E. S. W. *The Geology and Mineral Resources of Mauritius*, Vol 3 (HMSO, 1950)
SORNAY, P. DE. *La Canne à Sucre à l'Ile Maurice* (Paris, 1920)
THOMPSON, R. *Report on the Forests of Mauritius* (1880)
WALTER, A. *The Sugar Industry of Mauritius, a Study in Correlation* (1910)

BIBLIOGRAPHY

BOTANICAL

BAKER, J. G. *Flora of the Mauritius and the Seychelles* (1877)
BROUARD, N. R. *A History of Woods and Forests in Mauritius* (Port Louis, 1963)
HUBBARD, C. R. and VAUGHAN, R. E. *The Grasses of Mauritius and Rodrigues* (1940)

GENERAL ACCOUNTS OF THE ISLAND,
HISTORICAL AND MODERN

BACKHOUSE, JAMES. *A Narrative of a Visit to the Mauritius and South Africa* (1844)
BORY DE ST VINCENT, J. B. G. M. *Voyage dans les Quatre Principales Iles des Mers d'Afrique (1801 and 1802)*, 3 Vols (Paris, 1804)
BULPIN, T. V. *Islands in a Forgotten Sea* (South Africa, 1958)
DARWIN, CHARLES. *The Voyage of the Beagle*, Everyman edition
LABOURDONNAIS, MAHÉ DE, B. F. *Memoire des Iles de France et de Bourbon*, edited by Albert Lougnon and Auguste Toussaint (Paris, 1937)
LEGUAT, FRANÇOIS. *Voyages et Aventures*, 2 Vols. English edition and notes by Captain Passfield Olivier for the Hakluyt Society (1891)
MALIM, MICHAEL. *Island of the Swan* (1952)
Mauritius: Commerce, Industry and Tourism (Port Louis, 1970)
OMMANNEY, F. D. *The Shoals of Capricorn* (1952)
PIKE, NICHOLAS. *Sub Tropical Rambles in the Land of the Aphanapteryx* (1873)
Report on Mauritius 1966 (Port Louis, 1967)
SAINT PIERRE, B. DE. *Voyage à l'Ile de France, à l'Ile de Bourbon, au Cap de Bonne Esperance*, 2 Vols (Paris, 1773)

HISTORY

BARNWELL, P. J. *Visits and Despatches; Mauritius 1598–1948* (Port Louis, 1948)
BARNWELL, P. J. and TOUSSAINT, AUGUSTE. *Short History of Mauritius* (1949)
CREPIN, P. *Mahé de Labourdonnais* (Paris, 1922)
DRAPER-BOLTON, M. W. *Almanache de Maurice for 1854* (1954)

HILL, ERNESTINE. *My Love Must Wait*. Life of Matthew Flinders

HOLLINGWORTH, DEREK. *They Came to Mauritius* (1965)

LAGESSE, MARCELLE. *A la Découverte de l'Ile Maurice* (Port Louis, 1970)

LY TIO FANE, MADELEINE. *Mauritius and the Spice Trade. The Odyssey of Pierre Poivre* (Esclapon, Port Louis, 1958)

MACMILLAN, A. (ed). *Mauritius Illustrated* (1914)

OLMSTED, J. M. D. *Charles Edward Brown-Sequard* (Baltimore, 1946)

PITOT, ALBERT. *T'Eylandt Mauritius Esquisses Historiques 1598–1710* (Port Louis, 1905)

POPE HENNESSY, J. *Verandah, Some Episodes in the Crown Colonies 1867–1889*. Includes record of Sir John Pope Hennessy's governorship of Mauritius (1964)

PRIDHAM, CHARLES. *An Historical, Political and Statistical Account of Mauritius and Its Dependencies* (1849)

SORNAY, PIERRE DE. *Isle de France—Ile Maurice* (Port Louis, 1950)

TOUSSAINT, AUGUSTE. *History of the Indian Ocean* (1970)

——. *Harvest of the Sea. The Mauritius Story in Outline* (Port Louis, 1966)

——. *Une Cité Tropicale, Port Louis, Isle Maurice* (Paris, 1966)

——. (ed). *Early American Trade with Mauritius* (Mauritius Archives Publication, Escaplon, Port Louis)

TYACK, L. A. M. (ed). *Mauritius and Its Dependencies* and *The Seychelles, Treasure of the Indian Ocean* (Lausanne, 1965)

UNIENVILLE, J. R. M. DE. *Last Years of the Ile de France (1800–1814)* (Port Louis, 1959)

NOVELS AND LITERATURE ABOUT MAURITIUS AND BY MAURITIANS

CONRAD, JOSEPH. 'A Smile of Fortune', included in *Twixt Land and Sea* (1966 edition)

HART, ROBERT EDWARD. *Mer Indienne, Poems* (Port Louis, 1925)

LAGESSE, MARCELLE. *La Diligence s'éloigne a l'Aube* (Port Louis, 1958)

——. *Le Vingt Floreal au Matin* (Paris, 1960)

L'HOMME, LEOVILLE. *Poésies et Poèmes* (Port Louis, 1927)

ROSNAY, EDOUARD FROMET DE. 'Table Ovale-Poetes Créoles; anthologie Mauricienne/galerie poétique de l'Ile de France 1803–1897', *The Planters Gazette* (Port Louis, 1897)

SAINT PIERRE, BERNADIN DE. *Paul et Virginie* (first published 1788, often reprinted)

BIBLIOGRAPHY

POPULATION AND EDUCATION

BURTON, BENEDICT. *Indians in a Plural Society, A Report on Mauritius* (HMSO, 1962)
———. *Mauritius, Problems of a Plural Society* (1965)
HAZAREESINGH, K. *A History of Indians in Mauritius* (Port Louis, 1950)
NICHOLS, A. E. *A Report on Secondary Education in Mauritius* (Port Louis, 1949)
TITMUS, R. and ABEL-SMITH, B. *Social Policies and Population Growth in Mauritius*, Sessional paper No 6 (Port Louis, 1960)
WARD, W. E. F. *Report on Education in Mauritius* (Port Louis, 1952)

REFERENCE

Bibliography of Mauritius (1502–1954). This is updated by annual bibliographers (Mauritian Government Archives, 1954)
Dictionary of Mauritian Biography, edited by Dr Auguste Toussaint, published in twenty-five parts by the Société de L'Histoire de l'Ile Maurice (Port Louis)

TOURIST BOOKS

A to Z, Mauritius Tourist Guide
Guide Bleu, Madagascar, Comores, Reunion, Ile Maurice (Paris, 1955)
LENOIR, PHILIPPE. *Island in the Sun, Mauritius* (Port Louis, 1971)
WARD, ALEXANDER. *Climbing and Mountain Walking in Mauritius* (reprinted Port Louis, 1972)

ZOOLOGICAL

HACHISUKA, M. *The Dodo and Kindred Birds of the Mascarene Islands* (1953)
KEYNES, QUENTIN. 'Mauritius, Island of the Dodo', *National Geographical Magazine* (January 1956)
MEINERTZHAGEN, R. *On the Birds of Mauritius* (1912)
MICHEL, C. *Notre Faune* (Port Louis, 1966)
STRICKLAND, H. E. and MELVILLE, H. E. *The Dodo and Its Kindred* (1848)
WHEELER, DR J. F. G. and OMMANNEY, DR F. D. *Report on the Mauritius-Seychelles Fisheries Survey 1948–49* (HMSO, 1953)

ACKNOWLEDGEMENTS

As all visitors attest, Mauritius is an unusually friendly and hospitable island and I have, therefore, much kindness to acknowledge. I cannot thank by name the office boys and petrol-pump attendants who guided me when lost, waiters and room boys who served me well and talked of their hopes and ambitions, and many more who gave me time and information.

I can most warmly thank many others, and I especially appreciate the time and help given me by Sir Seewoosagur Ramgoolam, the Prime Minister; the hospitality of the Governor General, A. H. Osman; and the assistance and courtesy of Gaten Duval, Minister of External Affairs and Tourism; Sir Harold Walter, Minister of Health; the Right Reverend Edwin Curtis, Bishop of Mauritius; Peter Carter, British High Commissioner; Lt-Col and Mrs A. Ward; and Captain Ram of HMS *Mauritius*. I received also ready and welcome aid from Nuein Akaloo, MBC; Dr Adolphe, Government Archivist; Robert Antoine, director of MSIRI; André Beaumont; Philip Boullé of Rogers & Co and his staff, particularly Louis Espitalier-Noel and Marcelle Esclapon; Mr and Mrs Somdath Buckhory; Mrs Burrenchoby and the committee members of the Women's Self Help Association; André Cambier; André Decotter; Frederic Descroizilles; Mr and Mrs Leo Edgerley; Dr Fakim, editor of *The Star*; Regis Franchette and the staff of the Mauritius Tourist Office; Professor and Mrs M. E. L. Lim Fat; Jacques Giraud of Avis; Eddie Goldsmith, Comptroller of Le Reduit; Dr K. Hazareesingh of the Gandhi Institute; Guy Hugnin of the Mauritius Travel and

ACKNOWLEDGEMENTS

Tourist Bureau; Yves Lamargues, head of a Gaulette village; Marcelle Lagesse; A. Maingaard of Rogers & Co; C. Michel, Director of the Mauritius Institute; Captain Nicolin, director of Port Louis harbour; Claude Noel, President of the Chamber of Agriculture; Dr Alfred Orian, entomologist; A. W. Owadally, Conservator of Forests; J. Purmessur, Postmaster General; Mr and Mrs Yves St Flour; Harry Saminaden; R. Surdam, District Councillor, Surinam; Dr Auguste Toussaint; J. L. C. Vellin of the Government Development Corps; and Mohammid Vayid.

In London I should like to thank Cyril McGhee of Air France, I. D. Ramthoul of the Mauritius High Commission, John Sim of Martlet Travel and Allan Woolley of Houlder Brothers for all their assistance.

INDEX

Italic page numbers indicate illustrations

INDEX

INDEX